Also by Ryan Boudinot

*The Octopus Rises*
*Blueprints of the Afterlife*
*The Littlest Hitler*

# THE SHAPE OF REALITY TO COME

### RYAN BOUDINOT

ALEPHACTORY PRESS
SEATTLE

Alephactory Press
325 23rd Avenue East
Seattle, WA 98112

www.alephactory.com

First Alephactory paperback edition April 2018.

Interior Design by C.G.R.
Cover Design by C.G.R.

Library of Congress Control Number: 2018938445

ISBN 978-0-9994774-2-7
ISBN 978-0-9994774-3-4 (ebook)

# The Shape of Reality to Come

# CONTENTS

01    Introduction

07    Engineer the Renaissance, Part 1

17    Engineer the Renaissance, Part 2

25    Three Comments Overheard at the Evergreen State
      College Between the Years 1991 and 1995, Included
      Here for No Particular Reason Besides to Provide a Bit
      of Breathing Room Between Longer Essays

27    Seattle Invents the Future All Over Again

37    Ten Leadership Principles for the Hell of It

45    Socio-spatial Narrative

51    The Wave VR vs. 4DX

57    Why You Like Ready Player One

61    Virtual Reality vs. the Apocalypse

63    Culture = What You Value More than Money

67    What I've Learned from Playing Video Games

73    Immersive Media and Rural America

77    Invest Broad and Shallow, Not Narrow and Deep

81    Microsoft's Refreshing New Personality

85    Twenty Years in Seattle with Amazon

91    Four Essays on Artificial Intelligence

        I.     *AI-phobia*

        II.     *Will an AI Ever Learn to Love?*

        III.     *I'm Pretty Sure I Could Teach a Machine to Write a Novel*

        IV.     *The Highest Purpose of Artificial Intelligence is to Seek and Propagate Life Throughout the zUniverse*

103    Acknowledgements

105    About the Author

# INTRODUCTION

I first heard about virtual reality in 1989 or so, when I was a high school weirdo with a hideous car.[1] Nobody owned a laptop, and most computers weren't online. William Gibson's *Neuromancer* heralded a sexy new science fiction subgenre called *cyberpunk*, which explored inner spaces composed of information rather than outer spaces of planets and aliens. I happened upon an article in *Rolling Stone* about the fascinating, dreadlocked VR pioneer Jaron Lanier, and Gibson himself penned an article about VR in *CADalyst* magazine, one of the publications my father's civil engineering company kept on hand to bore potential clients. The arrival of the World Wide Web was still a few years out, but we were already dreaming of simulated worlds within vast seas of data. Then, abruptly it seemed, virtual reality fell off the cultural radar for the next quarter century or so.

In late winter 2016, I started to hear about an emerging story-telling medium that was largely being invented in my city, Seattle. Virtual reality, that science fiction vision of the late Reagan era, had crawled out of cultural hibernation. I soon fell in with a diverse community of geniuses and innovators who gathered at meetups and hackathons in warehouses and coworking spaces around the city. The

---

1    1977 Chevrolet Caprice Estate Classic Wagon, with a factory-installed 8-track player.

more of these friendly, funny pioneers I met, the more I wanted to contribute to the immersive media industry. That April, I attended the Microsoft Hololens Hackathon, a weekend that had me building a 3D holographic pop-up book with a team of brilliant Millennials. By August, I had signed up as an inaugural tenant of the University of Washington's VR startup incubator CoMotion Labs, alongside companies that were developing VR medical therapies, 360° films, and robot-fighting games. I had no clue what I was actually going to do there (not to mention how I'd pay the rent), but beneath my ambiguity and uncertainty I cultivated some faith that things would work out.

I wrote about Seattle's VR community on my blog, ryanboudinotisahack.com[2], approaching my subject like a small-town beat reporter, and soon the people I wrote about became friends and collaborators. I wrote an article for *Seattle Met* magazine that surveyed the city's VR scene as an emerging creative industry, approaching the piece as an exercise in self-fulfilling prophecy. A few months later, I founded a VR startup of my own that I named Starbird Reality, though, to be honest, I often feel it's more accurate to refer to myself as an "end-up" than a "startup." In November 2016, I started working as a consultant for Paul Hubert at Immersion Networks, a Redmond-based R&D lab.

As I write this, I'm a one-man company. I have no employees and no content to show for my efforts. The process of hustling for funding seems to involve a lot of bullshit posturing that I have little patience for. To be completely blunt, I don't even really know if I actually want to be a company. What I do know is that I'm obsessed about what immersion means for art and society, and how it might be a force for good in the world. As soon as I heard that immersive media represents new ways to tell stories, I started rolling up my sleeves.

One day when I was a science-fiction obsessed adolescent, a peculiar thought popped into my head. I decided my task wasn't to write science fiction stories. It was to write the kinds of stories that characters *in* science fiction novels would *read*. This book is largely an attempt to figure out how to tell stories in an age of shifting, unreliable realities. I

---

2    A domain that I was shocked to discover was available.

don't know what the hell I'm doing. I barely understand where I am. I am daily confounded by the political, environmental, and media landscape of my culture. Since 2015, I've been reading nonfiction exclusively, mostly books on science, technology, and innovation. As I worked on the book you're now holding, it slowly dawned on me with a mixture of horror and amusement that this is a business book.

Ha ha, suckers!

Starbird is the name of a road close to my childhood home in Skagit County, just south of Conway, Washington. When you stand on the Starbird Road exit and face north on a cloudless day, you're treated to one of the most inspiring views available in Washington State, the fertile patchwork of Skagit Valley with the San Juan Islands rising triumphantly in the background. My parents settled on seven acres there in 1974 and proceeded to raise me, my brother, my sister, a dozen sheep, and various other species of delicious livestock. The previous owners of the land had used parts of the property as a landfill; I spent my childhood playing amateur archeologist, extracting treasures—bottles, machine parts, agricultural tools—from the earth. I gathered these artifacts of the 1950s and '60s in an old refrigerated train car that I referred to as my "lab." When I hear the word Starbird, I return to the acreage where my imagination is rooted, where I enjoyed seemingly unlimited time and space to play.

An early memory. I'm just old enough to know how to read. I'm thumbing through a magazine and I come across an illustration of Lewis Carroll's Alice, her body composed of gears and circuits. The caption at the bottom reads "Alice in Technology Land." I ask my mother what this strange word means. She explains that it means machines and inventions. Immediately, a distinct, adult voice intrudes inside my consciousness like a visitor from another reality, and says, *Pay attention to this word, it's going to be really important to you.*

Ever since that moment, I've kept an eye on technology while developing my vocation as a writer. My father made his career as a civil engineer, and I had access to computers in the early 1980s at his office—sturdy HP, Digital, and Texas Instruments machines with monochromatic monitors wired to scritchety dot matrix printers. I wrote science fiction stories on those computers, storing my drafts on floppy discs. Years later, I enrolled in a graduate creative writing program and paid the bills by working as a customer service rep at Amazon, back when they only sold books. I always seemed to have one foot in literature, and one in technology. There have been many occasions when I've felt I don't fully belong in either camp. I've made peace with myself as a kind of hybrid, mutant freak, a writer who has twice been employed by a company that many believe drove a stake into the heart of literature, a tech industry professional who cares more about art than quarterly results.

The place where technology and art intersect we call media. My happiest moments as an undergraduate were when I was studying the history of radio, television, film, and recorded sound, gorging on mostly inscrutable theories of Frenchmen like Jean Baudrillard and Roland Barthes, American provocateur Neil Postman, and celebrated Canadian media theorist Marshall McLuhan. The fiction I was most drawn to belonged to the lineage of male, American postmodernists, starting with Thomas Pynchon, then Don DeLillo, David Foster Wallace, and George Saunders. In their works I found a thread—that the technologies through which we receive information fundamentally alter the nature of that information, and therefore alter *us* and our perceptions of reality. The music I listened to from my teens through my twenties—Public Enemy, Fugazi, Nirvana—questioned and deconstructed the bullshit, mainstream music we'd been force-fed. The cinema that attracted me was equally obsessed with authenticity, contrivance, and metafiction, whether in the ironic layers of Oliver Stone's *Natural Born Killers*, Spike Lee's fourth wall breaking, Richard Linklater's jettisoning the three-act narrative in *Slacker*, or the authenticity flip-flops of Orson Welles's *F for Fake*.

Countervailing these waves of Y-chromosome art and theory were the pissed off howls of my college town's Riot Grrrls; music from Sleater-Kinney, L7, Babes in Toyland, Liz Phair, PJ Harvey, among many others; and fabulist fiction writers of the nineties like Aimee Bender, Judy Budnitz, and Stacey Richter. The #metoo era in which I'm now writing feels familiar, like the amplified echo of the zines I read as a student in Olympia in the early nineties. I'll have more to say about the welcome intersection of the women's voices and immersive media in the pages to come.

I'm telling you this to fill you in on where I'm coming from. I belong to a generation inclined to climb on top of whatever artistic or theoretical movement occupies our attention at any given time to understand, contextualize, mimic, mock, prod, offset, praise, dismantle, spoof, and imitate it. This is how I think when I think about VR, AR, MR, XR, AI, and all the other coinages of the acronym wars. It seems weak sauce to just write about emerging, immersive media purely within the context of business models or the market potential of nifty new gadgets and games. I'm more fascinated with how the new media is changing our *souls*, man!

Will we ride these fancy distractions into an inescapable, fiery hellhole of unpreventable, ecological catastrophe and fascism? Or will they bind us together at precisely the moment that we need reminders of our common humanity, so that we can confidently confront the trials ahead together? I'm placing bets on the latter.

That said, I could very well be full of it. Wouldn't be the first time. Regardless, here are a few thoughts on how I organized this book and how you might get some use out of it.

I intend this to be an unconventional document that sparks readers' imaginations, leading to many more ideas than the handful presented herein. I imagine that this is the kind of book that you can cherry pick, opening it at random, shaking it once in awhile hoping that a stray thought or two falls out. Or maybe you can consult it when you're stuck; it would please me to no end if something in these pages pointed you toward your own ingenious solutions to creative problems.

When I started writing about virtual reality, the subject led me to other technologies, including artificial intelligence, video games, voice recognition, and immersive audio, and to places like a recording studio in the woods of Stanwood, Washington, one exit away from Starbird Road. When I think of the brilliant people I've met so far on this adventure, I'm reminded of an inspiring refrain from those long-ago Obama rallies: *we're the ones we've been waiting for.* The story of technology is animated by this spirit, and when this spirit flows among individuals, it binds them into a community. Like the gears and circuits that comprised Alice in Technology Land, our machines and media, when gazed upon from adequate height, reveal a shape that is identifiably human.

Seattle, March, 2018

# ENGINEER
# THE RENAISSANCE
# PART 1

So here's an admittedly unconventional business document that suggests that creative economies can be deliberately designed and cultivated. I've come to this conclusion while reflecting on the confluence of three of my formative experiences:

- Participating in the Pacific Northwest music scene of the late eighties and early nineties, in Skagit Valley, Olympia, and Seattle, as an aspiring musician and avid consumer of music.
- Helping launch Amazon's third-party seller platform in 1999.
- Leading an initiative, between 2013 and 2015, to get Seattle designated a UNESCO Creative City, part of a program designed to cultivate creative industries and connect them globally.

## Touch Me, I'm Sick

As a teenager in the Pacific Northwest of the late eighties, I had a bullseye on my head marked *GRUNGE*. I bought the 7-inch singles, wore the flannel, hurled my body enthusiastically into mosh pits. I started college in the fall of 1991 in Olympia, Washington, the birthplace of Nirvana, Sub Pop, K Records, and the Riot Grrrl movement, on the very day *Nevermind* came out. Nirvana released an album every year I was an undergrad, the arc of the band's success and tragic conclusion synced to my matriculation.

I showed up at school with a guitar, an amp, and a trove of pretentious ideas. I played my guitar and ~~sang~~ yelled in a band called Mugwump, wrote a bunch of songs, played shows. Music glued my social life together and was at the center of my friendships.

Even though Seattle became shorthand for this shaggy musical renaissance, the entire Pacific Northwest contributed to it, with micro-regional differences in style. Bellingham was known for pop and garage rock, typified by the Posies and the bands released by Estrus Records. Olympia was twee, feminist, and avant garde, home to Beat Happening, Witchy Poo, and Bikini Kill. In my hometown, Mount Vernon, metalhead farm kids absorbed punk and hardcore. Bands from Eastern Washington and Idaho also found fans on the Western side of the mountains. And, curiously, there was a bloodline connecting us to the other Washington, D.C., from which Discord Records sent bands like Fugazi and Nation of Ulysses to rock the all-ages clubs of the Evergreen State.

When I bore Millennials with yarns about the grunge years, I don't talk so much about the style of the music as the spirit in which it was created. It was an empowered, entrepreneurial, DIY time when we embraced self-sufficiency and restless experimentation. While a number of bands went on to great fame and success, the impetus for this music was organic and grew from the intensity of our fandom. Feeling cut off from much of the country in the years before the Web, various weirdoes found each other and committed to entertaining themselves and one another with music. We invested so much time, money, and energy in music because it was so fun.

Though I didn't realize it at the time, I was participating in a creative economy. I bought gear at music stores, CDs and vinyl at music stores, and tickets for shows at grange halls and community centers. I read *The Rocket* and *Backlash*—periodicals that advertised shows, reviewed albums, and interviewed bands. My bands rented practice spaces in storage facilities and occasionally made gas money performing. My friends and I were consumers and producers, musicians, entrepreneurs, zine editors, philosophers, fans.

Nobody asked for permission to create, consume, or market this music because there was no one around to ask. The gatekeepers— owners of record labels and venues, community college radio stations, zines, etc.—kept the barrier to entry fairly low, and encouraged anyone who had the wherewithal to become a gatekeeper him- or herself. Frustrated that your band isn't signed to a record label? No problem— just start a label of your own!

It's hard to overstate the pervasiveness of this empowered spirit, how it permeated our every endeavor, way beyond even music. I remember my friend and band mate Nate Manny, who later went on to play guitar in the Murder City Devils, getting into Ralph Waldo Emerson at Evergreen, finding a bridge between the self-reliance of American Transcendentalist thought and punk rock, particularly the Straight Edge movement. Nowadays he's co-owner of a branding and design firm. I'm convinced that the early nineties DIY ethos directly contributed to the entrepreneurial dotcom boom later that decade. The bands that blew up big were proof that our commitment to self-reliance could be a path to wide cultural relevance, which increased the level of participation, converted amateurs into professionals, and perpetuated an economy that took hold in mildewy basements, lofts, and trashed house parties packed to the gills with vomiting, tripping punks.

The key takeaway being that the grunge era was full of impassioned, talented, interdependent individuals who took it upon themselves to perform a variety of specialized tasks nobody had told them to take on, and most often didn't make money. This movement happened spontaneously, generated considerable wealth, and left a legacy that continues to enchant music fans the world over and draw tourists to the rainy Northwest. Hold that thought.

## zShops

In 1998, I started working as a Customer Service Representative for Amazon, which at the time just sold books. Within months of my manning the phones on the 2-11 pm shift, Amazon started selling CDs, VHS tapes (!), DVDs, and hardware. I'm of the mind that Amazon's most important innovation during this period wasn't expansion into a particular product line, but a reorientation to how the company fulfilled customer orders.

Around the time Jeff Bezos's grinning head appeared inside a box on the cover of *Time* magazine, eBay became ubiquitous. Many of us inside Amazon found their business model ingenious—no fulfillment center costs!—and soon Amazon set out to create its own, competing Auctions site. I volunteered to join the small team within Customer Service that helped launch this initiative.

Much of my responsibilities involved carefully explaining to customers why their orders wouldn't be shipped from one of the handful of Amazon distribution centers, and to try to calm them down when things went awry. This took awhile for customers to wrap their heads around, and there were plenty of auction mishaps that led to severely dissatisfied customers.

Complicating matters was an air of organizational arrogance fed by an enamored national media; I heard one Amazon VP boast that as soon as Amazon launched Auctions, eBay would be out of business within months. It didn't exactly turn out that way. Moments after the switch was flipped that allowed customers to upload content to the site, hardcore pornographic images started appearing on Earth's Biggest Bookstore. Oops. Of course, eBay continued to do brisk business, and Amazon had to confront the reality that the Auctions site had largely been a bust.

Rather than simply scrap the site, the company pivoted in a cunning and forward-thinking way. We'd built a system that allowed one customer to send something to another customer. Why did this system have to involve bidding at all? Why couldn't it feature products at fixed prices? From the ashes of Amazon Auctions, a new marketplace with an awkward name, zShops[1], was born.

By launching an auctions site, then converting it into a fixed price marketplace, Amazon seemed to be in the midst of an identity crisis. I remember a number of Amazonians questioning why we had decided to allow customers to take on fulfillment at all, endangering the customer-centric reliability they'd come to expect.

To reconcile these contradictory models, the company instituted

[1] I sure wish another name under consideration at the time had won out, by which I mean *Jeff Mart!*

the integrated detail page, which allowed third-party sellers to list their wares alongside the same items offered from Amazon's own fulfillment centers. To some, this seemed like madness. Why would we erode our own ecommerce business by allowing other sellers to offer the same items at lower prices?

The short answer was that, in terms of margin, Amazon made more on items sold by third-party sellers. The longer answer, I think, is that Jeff Bezos understood Amazon could make money by helping as many other people as possible make money. The old sell-pick-axes-and-shovels-during-the-gold-rush strategy.

What Amazon gained from all those individual sellers and buyers was better than a cut of their sales—it got a treasure trove of aggregate data. This would lead to Amazon's third transformation into Cloud services, which got underway during my second stint with the company around '04-'07.

Amazon wasn't just exceptionally effective at selling goods to customers—the company figured out how to leverage the power of its customer base *itself* to its customers. And it also understood just how much work customers themselves were willing to put into the site without being asked. Hold that thought.

### Creative Cities

In September, 2014, I represented Seattle at the UNESCO Creative Cities conference in Chengdu, China, among officials from over 50 cities worldwide who had come to discuss their cities' creative industries. I attended as the founder of Seattle City of Literature, a nonprofit that sought to get our city designated as a member of the UNESCO Creative Cities network. (Seattle was designated in 2017.) At the conference, I met delegates from Iraq, Uganda, Ireland, Australia, Mexico, Japan, and other countries, and discovered that we all faced the same, fundamental problem: How to get politicians to give a shit about art.

The purpose of the Creative Cities Network is to create opportunities for international collaboration centered around certain art forms and to spur the economic development of creative industries. Cities can be designated for literature, music, design, gastronomy, film, craft and folk art, or media art. Cities of literature can engage in writer exchanges, cities of music can host international festivals, and all of these various initiatives get promoted through a global network overseen by the UN's culture and education organization.

I learned that there can be many reasons why a city becomes renowned for a particular art form. Some cities, like Edinburgh, Scotland, claim a long list of canonical writers who have called it home. Other cities boast a commitment to arts education, or to civic programs that generously fund the arts. Some cities are distinguished by creative communities that gave rise to schools, movements, and masterpieces.

In the two years I led this project with Seattle's municipal government, the US State Department, and UNESCO, I learned about creative economies that encompass multiple, interdependent industries and the complex interplay of government, media, academia, and the marketplace as they relate to the dissemination of cultural products.

The case being made in that hotel ballroom in Chengdu was that investing in creative industries paid disproportionately big economic dividends. Cities that committed municipal resources to art saw increases in cultural tourism, and companies based in those cities had an easier time recruiting talent from outside the region if their city was perceived as favorable to creatives. Through this work I found my way to the writings on creative economies of Richard Florida and the theories of Jane Jacobs, and I came to appreciate the revitalizing, economic forces of creative enterprises.

I had seen this dynamic in action when I visited Reykjavik, Iceland, in 2011 and 2013. During my first visit, the country was still reeling, as if from a nasty brennivin hangover, after its economic collapse of 2008. I returned two years later on one of 50 airplanes destined for the Iceland Airwaves festival, in the midst of the country's recommitment to cultural initiatives and during the mayoral term of comedian Jon Gnarr. Now I saw cranes erecting buildings along Laugavegur Street. Most Icelanders I spoke to attributed this comeback to a spike in cultural tourism. In Seattle, ads for trips to Iceland had begun to wrap entire city buses, and a yearly festival called Reykjavik Calling brought musicians, chefs, and writers to Seattle for a program of performances.

As I was preparing this book for publication, I became aware of a recent study conducted by University of Kansas psychology professor Barbara A. Kerr. Dr. Kerr and a team of students conducted numerous interviews with Reykjavik's creative community to determine what factors led to the country's reputation for artistically punching above its weight. They concluded that the answer is an interplay of sociological, political, and interpersonal factors, including a pedagogical bias for allowing children unstructured play, egalitarian attitudes that put women's ambitions on equal footing with those of men, and an emphasis on making things by hand. In this country of a little over 300,000, it's also easier for aspiring enthusiasts to connect with world-renowned professionals, and the government—as I have witnessed firsthand—takes public art seriously.

When we think of creative hotbeds—Paris in the twenties, Haight Ashbury in the Summer of Love, Manhattan's Lower East Side in the seventies, Seattle in the early nineties—it's easy to fall into the trap of thinking that these cultural moments are lightning in a bottle, that they can't be predicted, and that they can't be deliberately instigated. We tend to believe that vibrant artistic scenes are the result of rare, visionary artists who suddenly appear among us in a flash, and we're lucky if we happen to be around when it happens. I disagree.

A renaissance doesn't happen by accident. It's the result of decisions made by self-motivated and socially supported artists and organizations who occupy specific roles and responsibilities. If you understand how these parties are interrelated and interdependent, you can drive policies that lead to more artists making better art (and getting paid better for it), resulting in more prosperous economies that benefit far more than the artists themselves.

Those thoughts you've been holding lead to three conclusions:

1. The self-motivation of individuals, which should never be underestimated, is amplified when these individuals are connected in a network.
2. Actively distributing influence and ownership through a network of many empowered stakeholders, rather than hoarding it for yourself, makes you massively more powerful.
3. A renaissance can be engineered.

### Engineering the Era of Immersive Media

For the past couple years I have participated in a new creative economy emerging in Seattle and various other places around the world. It's so new that it doesn't really even have a name yet, but more a salad of acronyms: VR, AR, MR, XR.[2]

As of this writing in the late winter of 2018, three companies dominate the nascent immersive media industry: Facebook/Oculus, HTC/Valve, and Magic Leap. Microsoft is also diligently advancing its mixed reality platform, and Apple appears to be invested in immersive media as well. These companies deserve a heck of a lot of credit for kicking off this particular wave of commercially available, affordable virtual reality. Facebook's $1 billion acquisition of Oculus and Magic Leap's teasers have sparked a great deal of enthusiasm for immersive media, inspiring a wave of creators, from software developers to musicians, to launch startups and connect with one another.

Adopting my Old Man Grunge persona[3], I often observe that Seattle's VR community these days feels an awful lot like Seattle in the summer of 1992. The self-motivated creative spirit is alive and hungry for expression. The community is supportive of newcomers and committed to continual experimentation and improvement as it becomes increasingly accomplished and professional. There are about 100 or so VR pioneers in Seattle that I regularly keep in touch with, and I've often thought that if, somehow, all these people were given an opportunity to work together, immersive media would quickly evolve in astonishing, world-changing ways.

In my opinion, the big three immersive media companies have made key strategic mistakes that will come back to haunt them. Facebook, marred by the intellectual and political immaturity of Oculus founder Palmer Luckey, erred early on by insisting on exclusive rights for experiences designed by indie developers. Valve countered this by offering non-exclusive distribution policies to developers, but allowed a glut of poorly merchandised, mediocre VR games to flood their store, Steam. This has been frustrating for indie developers, who are also chuffed by the $100 application fee to even be considered for

---

2    Virtual reality, augmented reality, mixed reality, and I forget what XR stands for.
3    This typically involves wearing a flannel shirt.

inclusion on Steam. Magic Leap, meanwhile, has teased the public from behind its ironclad wall of secrecy, training the very audience it should be engaging in the fine art of resentful schadenfreude. Few things make an indie VR developer more gleeful than the prospect of Magic Leap falling hard on its ass.

It seems to me that these companies are in a bind. We all know that virtual is going to be *a huge, game-changing opportunity of a lifetime* because, well, *it just is*, but nothing kills creativity faster than greed. Greed is just a compensatory reaction to the fear that one won't have the resources one needs to survive, and fear shuts down the parts of the brain that allow both rationality and creativity. At a recent VR talk I attended at an art gallery on Capitol Hill, I heard an Oculus employee explain that we just don't have the luxury of the time we need to craft great stories in VR. Behind this utterance I sensed the pressure of shareholders who picked on the theater kids in high school, business school know-it-alls who think that artists' flighty impulses need to be reined in, and Serious People who smirk at the term "liberal arts" and go to great lengths to educate the rest of us on how "the real world" works.

Obviously, I have something of a chip on my shoulder about how society regards artists.

Immersive media happens to be emergent while the tech industry at large goes through its tech bro reckoning with sexism. The macho attitudes that seem to erupt when you combine an MBA degree with a penis, such as a desire for *total market domination*, a propensity to mis-use the term "rock star," and a covetousness toward status symbols, contradict the properties of a medium that promises new—some would argue *feminine*—modes of empathy and understanding.

There is room for companies that focus relentlessly on what people want from immersion, companies that measure their success by how many independent artists they help to make an honest living. Organizations that are generous with their platforms and that have the humility to recognize they're participants in a more dynamic and interdependent creative economy. If there's one lesson that I learned from the grunge scene, Amazon's formative years, and a cultural diplomacy project that took me around the world, it's to not trust the loudmouth who claims to have all the answers. Instead, look for the passionate, self-reliant community that's asking the best questions. In the next essay, I'll pose some of those questions myself.

# ENGINEER THE RENAISSANCE PART 2

Late in the process of assembling this collection of essays, it occurred to me, that maybe I, of all people, have some practical advice to offer anyone who is obsessed with engineering a renaissance.

I don't have a prescription for turning your city into a creative hotbed. Every city has different challenges and is filled with different sets of individuals. But if you want to transform where you live into a place that makes a bigger commitment to creative endeavors, then I think it's necessary to establish new patterns of thinking and determine what values you want to champion. If you run a business and your definition of success is kissing shareholders' asses, well, the following is not for you.

I don't have answers; I have questions. These questions are going to be more effective if you sincerely embrace the desire to improve the lives of other people, champion innovation because it's way more fun than playing it safe, and, say, commit to human civilization's interplanetary survival for the next several thousand years. Once you have these convictions firmly in hand, the answers to these questions and thought experiments should become clearer.

Here goes...

**First, Take Stock**

Start by focusing on a particular creative product—book, song, video game, TV show, VR experience, whatever it may be.

Next, identify the cultural resources related to that product that are available in your city. Start a list, divided into three categories: people, places, and policies.

*People*

How many people in my community are empowered and self-motivated to contribute to this creative economy?

What specific roles do these people play?

How many roles are there?

How easy is it to fulfill more than one role?

How much money does one person need in order to produce one creative product?

How many hours does it take to produce a typical creative product?

What is the minimum viable number of people required to produce a creative product?

How many people make a living solely from creating said creative products?

How many people, given the chance, would choose to make their living solely from creating said creative product?

How screwed is someone who tries, and fails, to produce a creative product?

*Places*

Where do creators meet to exchange ideas?

What kinds of places are these? Schools, businesses, nonprofits?

Is there a predictable schedule to these gatherings?

Where do people go to experience creative products?

How many of these places are formal (businesses, or receiving funding from government) and how many are informal (peoples' homes, garages, street corners).

How many are public?

How many are private?
How easy is it to get a job in one of these places?

*Policies*

What media connects these people? Social media groups, blogs, meetups?

How is business contributing to the greater community, not just to their employees? Do they provide space for events, sponsorship, other forms of support?

How is academia contributing?

How is the government contributing? Divide this answer into symbolic contributions (plaques, arts districts, pronouncements) and actual contributions ($$$$). Ignore the symbolic contributions; they exist to trick constituents into thinking politicians are doing their jobs.

Are there policies that connect your creative economy with other creative economies, both nationally and internationally?

## Next, Identify Opportunities

Once you've created your list, you can get a sense of your city's strengths and how to identify areas of opportunity. You may find that your city has a lot of educational resources—classes at multiple education levels, facilities to host speakers and conferences—or maybe you find that there is an active community of artists who meet informally, but there are few school programs that support them.

Once you identify your opportunity areas, you can start to articulate goals to strengthen that sector of your creative economy. This usually means helping it make stronger connections to other sectors. Let's say you want to improve the relationship between your city's colleges and your indie film community. Then it's a matter of seeking out bored individuals who can make great collaborators when you promise them involvement in something cool. Filter for potential, not status. Find someone who works in the film studies program at a local college and ask if they'd be interested in hosting a night of locally produced short films. Institutions like colleges have spaces that often goes unused and tend to be institutionally inclined to engage communities. While my own experiences in academia have exposed me to some of the most maddening bureaucratic thinking I've ever encountered, education is still, broadly speaking, incentivized to embrace outside collaborators.

The more people involved in creating a single creative product, and the higher the cost of creating that product, the harder time you'll have establishing your city as a center of that product's production. That's why there's only one Hollywood, and only one Broadway. Movies, TV, and theater are expensive and require the coordination of a lot of people. It's not impossible to foster a creative economy for theater or film outside of LA and New York, it's just going to take longer.

Some of the best artistic advice I've ever heard came from a playwright named Mac Wellman, who spoke at a Bennington MFA residency I attended in the late nineties. Wellman said that whenever anyone asked him advice on how to make it in the theater business, he always told them to relocate to some city in the middle of nowhere, establish a theater company, and don't give up for 25 years. By that time, he said, you will become so good at producing plays that your city will become a destination for those who love the art form. Most people don't like advice that promises longer and harder work than what they're prepared for, but in my experience it beats advice that promises instant success for little effort, every single time.

## Holding Up a Mirror

As a Renaissance Engineer, part of your job is to hold up a mirror to your creative community. The article I wrote about Seattle's VR community for *Seattle Met* magazine, included in this book, was one of the most rewarding writing assignments I've ever had because I got an excuse to meet brilliant people working on new art forms, and I ended up making some friends in the process. I realized pretty early while researching that article that I wanted to serve this community by reflecting it back to itself. If I declared that Seattle was the place to be for VR, then there was a chance this declaration would become a self-fulfilling prophecy.

After the article came out, I heard from a woman who, upon reading it, quit her job and launched a VR startup. At least a couple people moved to Seattle to find work in VR after reading it. A local law firm specializing in immersive entertainment law saw their referrals triple after the article came out; one of the partners, upon meeting me, offered me some pro bono representation on the spot.

There is tremendous influence and responsibility in describing a creative community to itself. By acknowledging individual and collective effort, you provide self-motivated creators a context in which to calibrate their creative drives. And by context I really mean a narrative. We tell ourselves stories about ourselves all the time that dovetail with larger cultural stories, and we feel positive when these narratives align. For me, it felt reassuring to align my own humble musical efforts with the story of the grunge movement and my early professional career with the story of the dotcom boom.

In the year and a half or so when I was most actively blogging about Seattle's VR community for my sites ryanboudinotisahack.com and starbirdreality.com, I thought of myself as something of a small-town reporter. My readership wasn't very big—a few hundred people tops— but every one of those people was important, a pioneer. In the early stages of a medium, these are the most crucial people to find. Don't look for accomplishments or expertise; look for self-motivation, eclectic skill sets, and persistence. Look for people driven by curiosity who fearlessly jump into projects they're passionate about. One of your responsibilities as a Renaissance Engineer is to make it suck less when they fail.

## Figure Out What Metrics to Pay Attention To

You might be tempted to measure the success of a creative economy by how much money it makes. But as Hollywood demonstrates, you can have a creative industry that generates a lot of cash but still has deep, structural problems. I think it's useful, if more difficult, to define success by the prosperity of individual people.

Is the cost of producing a creative product going down?

Is the number of people engaged in producing creative products increasing?

Are there more opportunities for creators to learn, formally and otherwise, and to share their work with other creators?

Are creators incentivized or disincentivized to share knowledge with each other?

How many individuals occupy multiple roles?

How many artists dedicated solely to working in a particular art form can afford to buy an average house in your city?

Also, how much does it cost to completely blow it? What does failure cost?

The point isn't to eliminate failure. We all know by now that we shouldn't be afraid to fail. The point is to increase the number of failures while simultaneously driving down the cost of individual failures. This makes it easier to jump from one project to the next while increasing the opportunities to learn. Accept that you can't eliminate failure, but work to drive down the costs of those failures to the point that risks become attractive.

These are the kinds of questions we can ask to assess the health of a creative economy, not whether company X is making tons of money on a particular distribution device or company Y secured another round of funding. When we measure our success by the prosperity of individuals rather than the market ownership of a particular company or set of companies, we expand the size of the pie rather than immediately carve it up. If this sounds like hippy dippy bullshit, I'll happily point to the example of Amazon's Marketplace again. Remember, Amazon became gigantic in part because they figured out ingenious new ways for *other* people to make money on their platform.

### From individuals to groups

Once you're committed to the success of individuals, start looking at the successes of groups of individuals. Pay attention to your minimum viable creators number. Think about the size of these groups. Can they be smaller? The smaller the minimum number of creators necessary to produce a creative product, and the higher necessary investment they need to make, the more institutional support they tend to need.

Are your sectors (business, academia, media, informal gatherings) adequately connecting?

Are businesses providing internships to students enrolled in relevant areas of academic study?

Are innovations being codified into genres?

## Going International

Once you have your creative economy producing art at a healthy clip, consider the rest of the planet. The work you did to connect various sectors of your creative economy within your city will prepare you for connecting your creative economy as a whole to other creative economies around the world.

Start assembling dossiers on other cities, just as you created a cultural inventory of your own. You should notice two things right away. One, you'll find cities that are exceptional in certain sectors and lacking in others. When scouting other cities to forge alliances with, your ideal candidates will be exceptional at things your city lacks, and lack things at which your city is exceptional. We're talking about global trade on a cultural level.

Perhaps your city has exceptional educational programs devoted to an art form, but few businesses dedicated to producing that art. Find a city facing the opposite situation, and initiate contact between your universities and their startups. Initiate artist exchanges, and make sure your artists report their impressions, through the media, to their local creative community.

Much of the work involved in these suggestions, I might add, really boils down to finding email addresses and contacting someone with a courteous and well-crafted message. When I led the project to get Seattle designated a UNESCO City of Literature, I was astonished at how far I could get simply by emailing various creative industry figures in various cities around the world. It helps if you have a title, but I learned that oftentimes you don't even need that. Or, just give yourself a title. Sometimes the only qualifications you need to instigate change are being driven and giving a damn.

# THREE COMMENTS OVERHEARD AT THE EVERGREEN STATE COLLEGE BETWEEN THE YEARS 1991 AND 1995, INCLUDED HERE FOR NO PARTICULAR REASON BESIDES TO PROVIDE A BIT OF BREATHING ROOM BETWEEN LONGER ESSAYS

1. "One man's fetus is another man's canned food."

2. "I sure hope there's another war this year so we have something to protest."

3. "Dude, Dumpster diving is *so corporate.*"

# SEATTLE DISCOVERS PRESENCE

*An earlier version of this essay appeared, with a different title, in the September, 2016 issue of* Seattle Met *magazine.*

I'm swimming through crystals, navigating immersive color fields, and entangled in the towering vegetation of an alien planet. There is no up or down in this space as vast as a stadium. I'm also in artist Scott Bennett's comic book-filled garage studio in north Seattle, with a piece of consumer electronics strapped to my face. Bennett, who goes by Scobot when he VJs around town, pays the bills with gigs like painting sets for the Pacific Northwest Ballet, but in his spare time he's been exploring the artistic potential of the HTC Vive. It's in Bennett's studio that I encounter his abstract virtual reality art that he created with a program called Paint Lab. After an hour of immersion in Bennett's breathtaking 3D digital mind-scapes, I ask him if he has noticed anything different about himself since he discovered virtual reality. He pauses and gives the question some consideration. "In a lot of ways VR experiences feel like lucid dreams," he says, "I've actually noticed that my dreams are much more vivid now. It is possible that by using VR, my mind is developing some sort of 'muscle memory' and learning what it feels like to access that area of my brain where dreams take place. Who knows, but the real revolution of VR may be the unlocking of the human mind."

As you've likely heard by now, virtual reality (VR) refers to any experience involving immersive digital environments, typically viewed with a headset device that looks like a cross between scuba goggles and a cell phone. Augmented reality (AR) refers to a related experience in which digital elements are integrated with the real world you see before you. The recent Pokemon Go phenomenon, which had people scouring neighborhoods for Pikachu with their phones, is considered AR, but AR also includes experiences delivered via headsets like the Microsoft Hololens and the soon-to-be-released Magic Leap. The most popular VR headsets, the Oculus Rift and the Vive, come with hand-held controllers that allow you to navigate and manipulate various elements in virtual space. There's also the Google Cardboard and Samsung Gear, inexpensive devices which cleverly allow you to turn your smart phone into a stereoscopic display. Of course, we've been hearing about VR for decades, but now the technology has finally caught up to what used to be mere cyberpunk dreams.

Most of the companies responsible for VR and AR hardware have either established offices in Seattle or are about to. HTC, whose Vive is currently the most popular of the VR platforms, has an office in Pioneer Square and has formed a partnership with Bellevue game company Valve. VR company Oculus Rift, launched by inventor Palmer Luckey, was purchased for two billion bucks by Facebook, which opened a main Oculus office opposite the home plate corner of Safeco Field. Magic Leap, the secretive, Fort Lauderdale-based augmented reality company that tapped Vashon science fiction visionary Neal Stephenson as an advisor, recently announced plans to open an office in Georgetown in the Rainier Cold Storage building. Microsoft's AR device, the Hololens, is poised to go head to head against Magic Leap.

While these companies and others, including Amazon, Google, Apple, Samsung, and Sony, jockey to deliver the platforms and services that will allow us entrance to virtual worlds, a local ecosystem of creators is waking up to this powerful new medium's potential. All these new platforms are useless without content—apps, games, and whatever it is we're calling the VR equivalent of movies these days—most just call them "experiences." It so happens that a significant portion of this content, and the software used to create it, is being made in Seattle.

I've visited members of Seattle's indie VR community in their studios, offices, bedroom workshops, garages, co-working spaces, and at meetups and weekend hackathons. I've met game designers, educators, engineers, and artists building tools and applications that may very well disrupt every industry you can think of. Entertainment, social media, education, architecture, pornography, social justice, medicine, and the Internet itself stand to be transformed by this new technology. Seattle's pioneers of VR may just represent the emergence of the most powerful creative community this city has ever seen. And they happen to believe we're in for one hell of a ride.

It's after 1:00 am on a Saturday in May 2016 and I'll soon lose consciousness in Fremont Studios, a soundstage facility filled with game developers, cases of Red Bull, laptops, and Microsoft's Hololens, a wearable computer which superimposes 3D holograms over the real world.

Why am I here? The short answer: hell if I know.

Longer answer: I got invited to the Hololens Hackathon because I signed up after hearing about augmented reality as an emerging storytelling tool. Apparently Microsoft thinks it's valuable to include writers who don't know jack about code at these things.

As I lurch toward what appears to be a dog bed made for adult humans, I'm sideswiped by a seemingly inexhaustible 22-year-old named Eva Hoerth. Eva and I are on a four-person team developing a 3D holographic popup book called *Strange World*. We've recently realized that we can blow up our book to the size of a room. Our team leader is Majesta Vestal, a laser-focused recent Cornish grad and visual artist whose mother is younger than me. Doing most of the coding on our team is Tarik Merzouk, a recent UW Tacoma grad pushing himself to the limit of his understanding of Unity, the coding language used to create three-dimensional, immersive worlds. My role on our team appears to be to amuse my young teammates with tales of working as an Amazon Customer Service Rep during the Bill Clinton administration.

"Ryan! Come join our eighties dance party!" Eva exclaims.

"Jesus, Eva. Were you even *alive* in the eighties?"

"Nope!" Eva laughs and jogs over to a group of software developers preparing to embarrass themselves to the stylings of Axel-F.

Not only was I alive in the eighties, I remember hearing about virtual reality back then, too. *Holy crap*, I keep thinking, *it has finally arrived.* I collapse on the human dog bed next to a service door that bangs open and shut all night at irregular intervals, my brain converting holograms of sumo wrestlers and crime scenes into dreams. I've watched the Milky Way galaxy spin around me and I've poked around an archeological dig in Machu Pichu, all while wearing a device on my face that makes me look like an overzealous Daft Punk fan. The ostensible reason for this hackathon is to inspire developers to invent new applications for the Hololens. But it also feels as though we're the subject of a scientific experiment and that weeks from now, in some Redmond conference room, behavioral scientists will be puzzling over which snack foods make people code faster. I'm actually cool with that. It's worth being an augmented reality guinea pig if it means discovering a creative community that knows it's on the cusp of paradigm-shattering inventions.

After an interval of time in which I'm neither asleep nor awake, an Arcade Fire song compels me to roll out of the human dog bed. I stumble into the final stretch of this combination of Power Point throw-down, immersive geek-out, and the inauguration of a new medium. Our storybook app ends up winning an award for best visuals. Other teams unveil such apps as a room-scale architectural drafting tool, a cardiologist-training application that uses actual biometric data, a virtual garden, and apps that assist the blind by scanning the room for obstacles. When the 44 straight hours of coding and demos come to a close, I emerge blinking into the daylight with my commemorative Hololens T-shirt in hand, and it strikes me that once again, the future is being invented in Seattle. This time in Fremont, Center of the Universe, no less.

In *The Innovators*, a history of the digital age from Ada Lovelace to Google, Walter Isaacson pays considerable attention to the hippie ethos that birthed Silicon Valley as we know it. The communitarian and psychedelic backdrop of the late sixties and early seventies Bay Area intersected with advances in transistors to create a culture of innovation where Hewlett Packard, Atari, Apple, Google, and Facebook bloomed. Isaacson stresses that it's not enough for engineers to cook up technological innovations; the culture into which they're introduced has to be prepared to embrace, improve upon, and find imaginative uses for them. To say that Seattle's tech community is culturally poised to embrace VR is an understatement.

One of the most powerful forces animating much of Seattle's VR industry is feminism. A community of tech-savvy women—many of whom have endured some serious, grade-A workplace bullshit in the gender imbalanced video game industry—are working to ensure that Seattle's VR industry is structurally diverse and inclusive from the get-go.

Take, for example, the recent Women's VR Create-athon at UW hosted by the AR/VR Collective, a group that organizes community events and learning opportunities with the goal of expanding the industry's demographic scope beyond tech bros. Over a hundred women—developers, students, gamers—assembled in teams for a weekend at Startup Hall, where they made VR apps and took in lectures by female industry leaders. In contrast to the somewhat competitive spirit of the Hololens Hackathon, this gathering felt more mutually supportive and collaborative, more about learning and discovery than presenting and mansplaining.

Among the virtual wonders I sampled at the Create-athon was an application designed to train doctors how to resuscitate an infant, created by a team led by a medical student named Rachel Umorren. After donning the Oculus Rift headset, I was transported to an operating room where a virtual dummy of an infant lay before me. As the app quizzed me with a series of questions, I committed what can only be described as virtual malpractice on that poor dummy.

My Hololens Hackathon teammate Eva Hoerth was one of the Create-athon's primary organizers. "We had 100 women in the same room creating for VR," said Hoerth. "No one made a first-person shooter. I was so goddamn happy."

First-person shooters, of course, are what most of us tend to think of when we think of video games. I've sampled a few locally-grown first-person shooters in VR and had a blast every time. There's Invrse Studios' thrilling *The Nest*, in which you're a sniper picking off robots, Matt Matte's super fun skeleton-obliterating game, and Eric Nevala's atmospherically rich *Spellbound*, in which you're a wizard hurling fireballs at zombies in a tomb. These games, all under development as of this writing, are exquisitely designed, massively entertaining, and feature innovations destined to become industry standards. I had nothing but fun playing them and I'd love to play them again.

But what kinds of games can *only* be played in VR? VR facilitates heretofore unattainable sensations of empathy and intimacy in a digital environment; you experience a whole new magnitude of the willies when zombies encroach on your personal space. This feeling of *presence*, in which your brain is tricked into thinking you're actually there, is leading Seattle's game designers to ask deeper philosophical questions as they establish new game mechanics and themes. Rather than treating VR as an environment in which to grind out the same kinds of flat-screen shooters we've grown accustomed to, VR game designers are puzzling over what this new medium alone can do.

One indie game developer who's set up shop in a Sodo co-working space across the street from Oculus Rift's offices, Dr. Evie Powell, has been developing a VR snowball fight game that retains the thrill of combat in a more friendly, non-apocalyptic setting, a snowy field, surrounded by kids pelting you with snowballs. Not only can you create snowballs by "scooping" the snow with your Vive controllers, you can build snowmen that lurch off and attack your enemies. Dr. Powell is developing the game as part of the Oculus Rift Launchpad program, which provides developers a stipend as they create the games that will likely captivate us a couple Christmases from now.

Dr. Powell concedes, "I've had fun with my share of violent games. A good many of my favorite games from childhood were beat 'em ups, shmups [shoot-em-ups], and fighters. The psychological implications of violence in VR are just so much more concerning because VR players are so much more present. So when designers jump into the dark side of VR, I hope they seriously consider what players will take from the experience."

Dr. Powell stresses that it's not about banning or not developing violent games so much as it is about seizing the opportunity to create new kinds of games that only VR can deliver. "I hope that for every violent game that exists, there are other directions game designers go with their designs. This is one of the reasons why diversity of thought is so important. The last thing we need is for VR to be dominated by violent games. VR has so much more potential than just a platform for people that want to act out antisocial behavior."

All the VR game designers I've met have uniformly praised the community of people who have gravitated to this new medium. They share ideas and inspire each other, code all night together, and embark on cannabis-enhanced sessions of Tilt Brush, a popular 3D drawing program for the Vive. The empathy inspired by this new medium seems to encourage these creators to themselves practice more empathy, and to collaborate, share, and socialize.

Tom Doyle, formerly of game company Bungie, runs Endeavor One, a bootstrapped startup that I was surprised to learn is headquartered a block from my apartment on Capitol Hill. Turns out that for years I've been scowling at the same Red Box as the Lead Artist for *Halo*, whose startup exclusively develops games for VR.

In Doyle's garage I slip on a Vive headset and settle into *Duel*, a vivid playing environment that feels like *Miami Vice* made a baby with *Tron*. Tom encourages me to knock down floating cubes with a virtual bow and arrow, a blaster, and grenades. After a round of virtual target practice among some of the richest visuals I've seen in any Vive demo, I pull off the headset and blink. I'm back in the garage, standing beside Tom's Subaru. Or, to put it another way, I never really left the garage, even though it feels like I did.

Doyle appreciates the community spirit animating Seattle's VR gamer scene. "Never before in my sixteen-year career have I seen so many teams doing their best to help each other out as much as possible. So many of the walls that divided this type of talent, historically, are being torn down in this fledgling industry," Doyle says.

Among the walls being torn down are those between VR technology and education. Lisa Castaneda is the founder and CEO of Foundry10, a north Lake Union company that provides technology-based enrichment opportunities for schools. Foundry10 introduces schools to VR tech then studies how they use the technology. This past year, Foundry10's VR pilot program has expanded to include not only local schools but schools in Hawaii, Toronto, and Finland, among other places. Castaneda buzzes with the enthusiasm of someone who can't wait to tell you about the Next Big Thing, and regularly drops references to the latest academic research and inserts terms like "cognitive load" into conversation.

After she launched the VR program, Castaneda observed that kids don't just want to consume experiences in VR. They want to use VR to design experiences of their own.

Says Castaneda, "Some schools come into it with an idea that it will be used in a game design course, or computer science class. Then there is the emerging content creator group, where kids become so inspired or challenged that they begin to work in game engines like Unity and Unreal to create their own content."

Castaneda and many of the other VR innovators I spoke to are wrestling with questions of how this powerful medium will change individual human psychology and society, all while they invent the very applications provoking those changes in the first place.

A bit up the hill from Foundry10, in a house near 50th and Latona, two recent UW graduate students, Nick Connell and Todd Little, demo their VR language learning app in a bedroom converted into a Vive space. I strap on the Vive and find myself behind the counter at a café, where I'm asked to prepare a variety of foods in response to orders in French. The app is definitely in a pre-beta phase, but I can already tell how useful it will be to students who want to learn another language.

"VR is a new medium to explain ideas," Connell says after the demo, over a pleasant array of wine and cheese, "What you're going to start seeing in VR are certain things you can't accomplish in other media. Presence will improve retention."

But will VR just make us all homebound, plugged into devices and even further isolated from real people? This is the kind of question that occupies Troy Hewitt, the Co-founder and Director of Communications and Connections at a Kirkland-based startup called UGen. It's in a conference room devoid of furniture at UGen's offices that I play a virtual game of capture the flag against opponents who control their avatars in the next room on PCs. It's a blocky, Super Mario kind of experience for now, with greater texture and functionality to come. UGen's mission is to create virtual spaces where users can create their own games then invite their friends to play in them together. It's multiplayer social gaming in customizable worlds, and it's precisely as fucking awesome as that sounds. After I launch myself into the sky using a jetpack, I realize that I have to come back down. As the ground rushes up to meet me, my real-life knees tremble and my body braces for physical impact that never comes. I pull off my headset and blink at a nearby whiteboard covered in UX terminology.

"We're building a place to build an experience," says Hewitt, underscoring the notion that VR is going to be another way to interact with people, albeit in avatar form.

One place to interact IRL (in real life) with Seattle's VR pioneers is at CNDY Factory, a multimedia production facility and community gathering space on Dexter north of Mercer. The brainchild of new media expert Tim Reha, CNDY Factory occupies a building that in previous lives has been a literal candy factory, jazz recording studio, and a clandestine pot growing operation. Word has it there's a fully equipped SM dungeon in the basement. If there ever was a perfect epicenter from which to launch a media revolution, this has to be it. Case in point—in July, CNDY Factory hosted a gathering of local VR leaders and the VR Venture Capital Alliance, an organization representing $2 billion in funds for VR/AR startups. The conversations happening at CNDY Factory are already leading to new projects, startups, and alliances.

Reha wants to establish Seattle as not just a place where big VR hardware and software companies vacuum up local tech talent, but as a major VR content creation center that can go toe to toe with Hollywood. As film studios in Los Angeles launch their VR divisions, Reha and others who orbit CNDY Factory are figuring out how to convert the energy of a community that's bursting with ideas into content production. There's no reason, Reha believes, that we should assume LA alone will deliver the cinematic VR experiences of the future.

"We're in the rain shadow of these big hardware companies, and no one is writing checks for content," Reha says.

Maybe so, but places like CNDY Factory are attracting the talent that is producing the next wave of VR apps and experiences. It's at a

CNDY Factory VJ night where video artists are projecting their work on the walls that I meet a young musician named Gus McManus who is demoing his new VR DJ mixing tool. I strap on an Oculus Rift headset modified with a Leap Motion sensor, which converts my hands in front of me into robotic hands in VR. Suddenly I occupy a massive DJ booth, surrounded by a variety of buttons and faders. As I manipulate these controls in virtual space, the real space of the CNDY Factory throbs with the loops and samples I trigger.

After I pull off the headset and hand it to the next DJ, I stand alone on CNDY Factory's deck and take in a late night view of Lake Union. I feel fortunate to exist at precisely this time, in precisely this place, witnessing the birth of such a powerful new medium. The pioneers of virtual reality have arrived, and to a person they sense the magnitude of what's coming. The city that invents the future just discovered presence.

# TEN LEADERSHIP PRINCIPLES FOR THE HELL OF IT

*First Principle: Knowing Who You Are is More Important Than Knowing What You're Doing*

We enjoy being considered authorities in the things we're passionate about. Authorities have things under control. They know what they're doing. That is, until everything goes to hell. That's when authorities get tested and they either rise to the occasion or prove undeserving of their authority in the first place.

We—especially men—have been conditioned to believe it's more important to appear in control than to reveal uncertainty or vulnerability. But these ambiguous, uncomfortable conditions are the places from which we grow. Great artists and leaders purposely put themselves in positions of uncertainty and fear. When I taught creative writing, I often told my students that feeling stupid and scared was a sign of growth; such a state just means you're working at the limit of your intelligence and ambition.

The trade-off to operating in a state of vulnerability and uncertainty is that you learn more about yourself than when you play it safe. The more you understand who you are, the easier it will be to make decisions when you're faced with no obvious solution. You use the results of those decisions to better understand yourself.

The immersive media industry is at such an early point in its history that the unknowns far outnumber tried-and-true methodologies. This means it's incumbent upon you as a creator to gaze inward, interrogate your talents and biases, and mindfully curate your own taste and sense of self. Expertise doesn't mean showing off what you know. It means finding intelligence in neuroplasticity and courage in understanding your own nature with ever greater clarity.

*Second Principle: Be Obsessed with Potential, Not Status*

The first thing Genghis Khan did whenever he sacked a city was round up the aristocrats and lop off their heads. Everybody else got a job interview.

One might argue that the Mongol empire grew as rapidly and vast as it did in part because they observed this extreme form of meritocracy. You were judged by how useful you could be rather than by how other people regarded you.

Status can blind us to the potential and value of other people. In "Positively 4th Street," Bob Dylan sang:

> You've got a lotta nerve to say you are my friend
> When I was down you just stood there grinnin'
> You've got a lotta nerve to say you got a helping hand to lend
> You just want to be on the side that's winnin'

Most of us have had fair-weather friends like that. Strivers and climbers who drop us the moment they perceive that our social standing has slipped. In this way, life can feel like a giant game of chutes and ladders, of jockeying for position on the heap. Being a dismissive snob to those perceived as beneath you is almost always paired with an attitude of reverence for those perceived as above you. These feelings are like a muscle group of status, and they can be severely limiting when operating within the context of an emerging medium. To eliminate status and open yourself up to potential, resist indulging in worshipping big things just because they're big, and dismissing small things just because they're small. What's small today could be huge tomorrow, and everything that is huge today was once small and could easily, and quickly, disappear tomorrow.

If you resist the status game, you begin to better perceive the potential of other people. What's most magical is that the very act of recognizing potential is often the necessary trigger to bring that potential to fruition.

*Third Principle: Maintain Equilibrium Between Creative Freedom and Artistic Excellence*

We trust artists. Trusting artists means staying out of their way and respecting the creative process, which can become messy and rife with blind alleys, failed experiments, and uncertainty. Trust means not forcing artists to repeat themselves or imitate another artist just because something succeeded in the past.

And it means cultivating an environment where artists are not afraid to provoke or offend. Social media gave us an era of public shaming, trolling, and "stay in your lane" rules devised by self-appointed social norm police. Not only do we reject these limitations on expression, we aim to destroy them with the empathetic generosity of the new media.

We proceed with vigilant understanding, assume good intent, and default to giving those with whom we disagree the benefit of the doubt.

We're pro-masterpiece. We endeavor to transcend the social conditions of the moment by creating medium-defining art that will resonate for centuries and ensure humanity's evolution beyond earth. We believe that accepting mediocrity in an attempt to spare someone's feelings equals artistic death. Our considerate and unvarnished honesty is of a piece with our kindness.

Over-indulgence of creative freedom can lead to diminishing artistic standards and a hobbyist-level of commitment to artistic production.

Over-emphasis on status-based signifiers of artistic excellence can quickly squelch creativity and trick us into equating market success or ephemeral critical approval with artistic value.

We balance creative freedom and artistic excellence in order to elevate the art that will long resonate with the people of the future.

*Fourth Principle: Compete Like Artists, Not Athletes*

Reflect on how John Lennon and Paul McCartney wrote "Penny Lane" and "Strawberry Fields Forever." Having established themselves as the world's greatest pop song writing duo, Lennon and McCartney kept pushing each other to go further. Paul delivered "Penny Lane," which vividly imagines a neighborhood from his childhood, teeming with absurd characters and inside jokes. The song is both epic and buoyant, expansive in scope and quotidian in detail, a true masterpiece. John must have felt both inspired and challenged, as he responded by writing "Strawberry Fields Forever," a song equally rooted in the past, but committed to the discombobulating interior landscape of childhood, rich with quantum states of confident doubt and sublime transcendence.

It's been said that Lennon and McCartney competed with one another like someone climbing the rungs of a ladder. First one would advance, then the other. They spent their careers inspiring each other then besting each other. This seems like an instructive metaphor for how we might grow the immersive media industry.

I often wonder if capitalism operates the way it does because most of the people in charge are white men raised on team sports and fraternities. While many of my peers spent high school playing sports, I was practicing with my band and figuring out how to get booked at parties and clubs. There was definitely a competitive aspect to playing in a band, with all gradients of shade thrown at bands whose popularity surpassed ours. But in general, the local bands that displayed true talent inspired me and pushed me to play better.

The world is full of people who perceive the economy as a relentless, zero-sum scramble for market dominance and endless conflict. But when an industry is in its infancy, individual competitors work against their own self-interest if they crush all the competition. It's like a single carrot in a garden deciding to grow big by killing all the tomatoes. When you're growing a whole new sector of the economy, it's more important to cultivate a customer base as a whole than to carve whatever base there is into portions.

If the various startups and individuals that comprise the emerging immersive media industry are in competition with one another, I would hope it is in the spirit of a bunch of bands that share the same handful of venues and record labels, not corporations ramming their office towers into each other as in that classic Monty Python sketch.[1] We understand that no single company or idea will succeed unless success is broadly distributed. We're looking to each other for inspiration to bring new masterpieces to fruition so that we can rise together.

*Fifth Principle: We're a Feminist, Queer, and Multicultural Organization*

In a better world, we wouldn't have to spell out this principle because it would be self-evident. But here we are in the Trump era, and work remains to include and respect women, people of color, people of various faiths, and LGBTQ people at every level of human affairs. We recognize that systemic problems require systemic solutions, and are committed to working for positive change.

We favor concrete, constructive action designed to empower the less privileged over symbolic, negative action designed to shame and humiliate those with whom we disagree. We constantly investigate whether our efforts for inclusion result in tangible results, or if they're just moralistic window dressing designed to flatter our sense of self-righteousness on our feeds.

Immersive media belongs to all humanity. We believe that the talents and leadership of women, people of color, and LGBTQ people are necessary for its success, full stop.

*Sixth Principle: Refer to Principles Before Relying on Plans*

The delight of discovery far outweighs the reassurance of sticking with a pre-ordained plan. Artistry requires the momentum of confidence in the midst of ambiguity.

We craft plans in order to point ourselves in the right direction, establish our accountability, and make sure we have all the resources

---

1    "The Crimson Permanent Assurance."

we need, but not to anticipate all problems before we've encountered them. Pioneering a medium that's in its infancy means that there's simply no way to prepare for every contingency.

Plans are counter-productive when they protect us from having to do the necessary work and make decisions. Principles help us discover, within ourselves and collectively, the path forward.

Don't brace against uncertainty, embrace it.


*Seventh Principle: Bias for Action*

This principle is straight-up ripped off from Amazon. (Thanks, Jeff!) As stated in Amazon's "Leadership Principles" document:

> Speed matters in business. Many decisions and actions are reversible and do not need extensive study. We value calculated risk taking.

When most people hesitate to take decisive action, it's out of fear that they'll get in trouble. And true, sometimes decisive action can screw things up for everybody. But that's the necessary cost of innovation.

The opposite of a bias for action is stagnation, and stagnation is untenable and unacceptable. It's incumbent upon us to be forgiving of mistakes that were motivated by calculated risk taking, and to encourage the occasional leap into the unknown.

Never wait for the ideal conditions to start something. Ideal conditions are a myth and constant motion is crucial. If you're stuck waiting for a green light from somebody, work on something else that amuses or delights you in the meantime. Avoid becoming the writer in Henry Miller's *Tropic of Cancer* who so idolizes Fyodor Dostoevsky that he's too paralyzed to start a novel of his own.

Make what's just beyond possible using the resources you have, like Miles Davis's high school music teacher instructing him to play a step above his own ability.

If all you have are two popsicle sticks, make the coolest thing that's humanly possible to make with two popsicle sticks.

*Eighth Principle: Deadlines are Sacred*

They don't just temporalize our productivity, they frame our existence. We're on a deadline towards our apparent departure from physical reality. Even reality is on a deadline toward the heat death of the universe. You're going to be dead sooner than you think. Might as well finish that project.

*Ninth Principle: Exaptation*

When Tom Waits was asked what he did with his song ideas that didn't become songs, he replied, "I cut 'em up and use 'em for bait."

Use the leftover pieces of failed experiments and build something else. Be cognizant of byproducts and be receptive to turning them into products. Try using something designed for a specific purpose for a different purpose. Discover the secret potential of something by placing it in a new context. Take two ideas that seem to have nothing to do with each other and see what happens when you force them to play together.

*Tenth Principle: A Useful Toy is the Same as a Fun Tool*

Elevate the concept of fun so that it is equal to the concept of usefulness. Resist the Puritanical conditioning that tells you that fun things are frivolous. Art depends on the cultivation of bliss. To work in a state of joy is itself the highest achievement. Making someone else laugh, cry, or reflect with art is an act of moral generosity.

The emotional richness of your life and your refined taste is your power. The fact that you have people who you love and choose to spend time with is not an inconvenience, it's integral to what makes you an appealing human being. Our worth is derived from the quality of what we produce, not the hours we're seen by a boss sitting at our desks.

Demolish the distinction between work and play. Toys have a purpose. The best tools are fun to use.

When you're doing your life's work, you just call it living.

# SOCIO-SPACIAL
# NARRATIVE

I.

When you ask a typical person what a story is, he or she is apt to say something along the lines that a story has a beginning, middle, and end, or that it involves a hero and antagonist. These can certainly be attributes of a story, but I'm more inclined to probe this question by meditating on the purpose of story. What does a story actually accomplish?

One way to think of a story is as an aesthetic pleasure-delivering device. If you consider a story in the abstract, divorcing it from its content, you behold a system that regulates curiosity and enlightenment. Stories are about problems and exceptions to everyday life, and typically resolve by leaving the characters and readers with renewed understanding of their world. James Joyce, in his story collection *Dubliners*, popularized the concept of the epiphany, or moment of truth. We are attracted to stories because they're little machines that start with a problem and end with a character surmounting that problem and learning something important about him or herself in the process.[1]

---

1    This definition purposefully ignores the late twentieth century flowering of post-modern literary forms (Oulipo, Dadaist word collages, Burroughs's cut-ups, whatever you want to call Kathy Acker's work, particular strains of patience-testing prose poetry,

The best storytellers are masters of regulating this push-pull process of curiosity and epiphany. The *Harry Potter* series, on one level, is about a school for young wizards. On a more abstract level, it is about the deft introduction and resolution of elements and tensions. I'm convinced that JK Rowling's popularity is a result of her mastery over conflict, novelty, and epiphany, rather than the appeal of child wizards, per se.

When we speak about narrative arcs, we might consider that the narrative arc model, popularized on workshop white boards with Freytag's triangle, is itself a metaphor. An arc is the parabola of a rocket's trajectory. An arc is a rising and falling. We think about narratives as operating in relation to the earth's gravity. Further, we think of an arc in two dimensions, a line that proceeds from left to right, mimicking the trajectory of words in Western languages on a page.

The spatial properties of virtual reality mean that the old metaphor of the narrative arc is no longer sufficient. We're struggling to shape storytelling in virtual reality in part because we're applying old narrative models in a medium with unexplored properties. It's like trying to use flat head screwdrivers on phillips head screws.

A narrative in virtual reality still operates according to gravity, but that gravity is like a star acting on other celestial bodies, rather than a planet acting on a figure stuck to its surface. Instead of rising and falling, a story in VR is pulled toward and pushed away from objects, an interplay of interiors and exteriors, of penetration and emergence.

If we take to heart McLuhan's idea that the content of any new medium is old media, then we can see ample ways that traditional, linear narratives can be embedded within VR. One idea—old narrative methodologies can exist in virtual reality if they are presented via representations of the old media from which they came. Perhaps there are television sets in VR that broadcast shows, or books that users can pick up and read. The structural properties of old media can't govern the structural properties of VR, but can exist as *content* within VR. The membrane of virtual reality fully encompasses old media.

But it is so much more than that. Imagine how the social elements of VR influence story. The emergence of such social VR experiences as The Wave, Altspace, and Rec Room suggest new opportunities for

---

etc.) that claim the label of "story" but deviate from expected norms of conflict and resolution. Consider these exceptions that prove the rule.

storytelling the likes of which we have never enjoyed as creators. Social VR is the first art form that includes the viewer inside the work of art itself, indeed converts the viewer into a participant, co-creator, and aesthetic element of that experience.

## II.

The narrative unit of fiction is the chapter. The narrative unit of movies is the scene. The narrative unit of VR is the realm.

The transitional mechanism of a work of fiction is the chapter break. The transitional mechanism of a movie is the cut. The transitional mechanism of a VR realm is the portal.

Portals connect realms.

## III.

The state of social VR as of this writing is simple. Build a realm that people can hang out and do stuff in together. The realms I've experienced so far are beautiful and engaging, but sometimes I'm at a loss as to what I should be doing in them. This gets me thinking about summer camp. A summer camp isn't a free for-all. There is a schedule, a social framework. It's temporal. Part of the pleasure of summer camp is the interplay between free play and counselor-led activities.

The same principle can work in VR. We're seeing a bit of this dynamic at work in such things as events, like The Wave's Wednesday DJ performances.

A schedule in VR can include such things as the appearance of gigantic pit in the ground at 3:00, a passing storm, or a limited opportunity for everybody's avatar to get a new head. This schedule serves as a narrative structural element.

A review our structural elements: Realms. Portals. Schedules. A realm is a place. A portal is a link between places. A schedule is the temporal blueprint for what happens in a realm. The artistic experience depends upon the interplay between these elements and the participant's free will.

IV.

Picture a series of realms as a solar system orbiting a sun. The narrative gravity pulls you from the outer realms, closer to the sun's nucleus.

The experience opens when you appear in an outer realm, alone. You interact with the environment and gather information. Perhaps you're alone in a room with a TV set broadcasting a story. Maybe you piece together what happened to this place based on notes, audio clips, or other ephemera.

At some point you discover the portal in the form of a lion's mouth. You let the lion eat you and appear in a new realm. In this realm you meet another participant who is signed in via their own home VR rig, elsewhere on earth. You exchange information about the realms you just came from. This other participant might even be an actor or AI employed to perform certain actions or deliver specific lines. The story depends on your sharing and withholding information from each other. Is this new person a rival or an ally? Do they want the same things you do? The story charges on or stalls depending on what information is shared or withheld.

Every portal leads to a new realm that is more populated with player-characters than the last. The narrative nucleus, or the "sun" in our metaphor, is a realm where all player-characters converge. The coming-together of all the player-characters ties into the thematic climax of the story.

In this way, we can convert players into characters. Players are both witnesses and participants. This radical aesthetic shift could represent a sea change in how art is created and perceived.


V.

Trust aesthetic experience.

When I taught creative writing, students sometimes asked me how many pages a novel or short story should be. My answer was that a story should be precisely as long as it takes to tell it. It's a matter of paying close attention to what a particular story wants to be, rather than imposing external structures upon it. On the other hand, our stories accommodate externally defined structures all the time. The episodic novels of Charles Dickens largely conformed to the publishing schedules of London's newspapers. Stories told via television evolved to accommodate the economic necessity of stepping out of the story at specific intervals to advertise products.

These structures will emerge in virtual reality just as they have for the novel, the stage, and the screens. Genres will emerge and tropes will be established. The most important question at this stage of immersive narrative is, what do the stories want from us? What delights and compels us and our audience? What stories are only possible to tell with the media of presence and immersion?

# THE WAVE VR
# VS. 4DX

The Wave VR will be spoken of a hundred years from now as an early, formative masterpiece of its medium, like George Millies's *Le Voyage Dans le Lune* or Chuck Berry's "Johnny B. Goode." I'd heard about The Wave as a VR music experience, and early video clips made it out to be some sort of toy for EDM kids. But after spending an hour or so with it on a Saturday, I'm happy to report it's much, much more than that.

The Wave is a social environment stocked with music, toys, and trips. There's a candle-lit cave in which you discover a DJ booth, and a vast, subterranean plateau where various objects hover and glow. You are represented as an avatar and can choose a number of different heads including a panda bear and a wicked-looking helmet. You teleport using the thumb pad on the Vive, and when other visitors teleport, their movements resemble those of stones skipping over water.

The visuals are trippy and hint at just how weird virtual environments are going to become. With "money" that you earn by grabbing hovering orbs, you can buy "trips" and "toys." Toys are things like guns that shoot musical projectiles. Trips are nested environments that you can share with other visitors. Some trips are more representational, like the one where you're a fish with a head-mounted probe that triggers a balloon drop when you touch it to the probe of another fish. Others are more abstract, like one where you sort of become liquid. And then there's the trip of the dancing television-headed people. Much of the pleasure of trips is that you share them with other visitors, and this social element is what makes The Wave so compelling.

When I sidled up to a dozen or so avatars hanging out by the trip and toy station, I felt as though I'd walked in on some sort of party in progress. I heard voices—a man's British accent, someone speaking French, a Japanese speaker, a dude who sounded stoned ("This shit is definitely 420-perfect!"). Then I heard a child's voice and turned to see an avatar that was shorter than the others.

"Hey!" the kid said, "How does this work? How do I get money? Someone help me!"

I went into Dad mode and approached the kid's avatar and tried to coach him as well as I could, aware that my avatar resembled a ring wraith from *Lord of the Rings*. I felt protective, especially as avatars around us were exclaiming things like "Holy fucking shit!"

The English gentleman seemed to understand how the environment worked more than most, so I sidled up to the Brit's avatar and became friends by offering him a friendship orb. He invited me to his home cave, where he played music on his DJ rig while I controlled the visuals, manipulating the pulsating geometric animations that surrounded us as if we were under a dome.

I realize that translating this VR experience into language makes it sound utterly insane. I'm having to use the kind of words I'd use to describe a dream. I've been blown away by VR environments before, whether in Tilt Brush, The Nest, or Google Earth. And I've tried social VR, sampling VREAL and Envelop VR's platforms, for instance. The Wave takes what's so compelling about VR and pulls it together into an explorable realm. I constantly found myself thinking oh, wow, so it can be like THIS.

When a Japanese speaker's avatar held out an object to me and said "It make music," I was awestruck by being connected to an abstract representation of another human being in an imaginary world despite our vast geographic separation.

The Wave suggests a genre I've decided to call the *realm*, the closest thing yet to the Oasis of Ernest Cline's *Ready Player One*. This new format feels about on par, in terms of a media paradigm shift, with the invention of movies or websites.

Speaking of movies, the same week I experienced The Wave, I also saw *Fate of the Furious* at Seattle's Regal 16, a theater that recently upgraded a number of its auditoriums with plush recliners. While I was apparently not paying attention, the theater added something called "4DX," which means your seat physically moves and you occasionally get mist sprayed in your face. Imagine getting alternately fanned and spritzed while sitting on a washing machine in spin cycle while a toddler kicks you in the back during an earthquake. It's exactly as fun as that.

About ten minutes into the movie I was prepared to leave and ask for a refund, but then I realized that this experience would be great fodder for an essay, goddamn it.

Enduring the 4DX *Fate of the Furious* right after having my mind blown by The Wave exposed a yawning chasm between movies and VR. What pissed me off about 4DX was that I had no agency whatsoever. I was expected to passively accept the jolts and rumbles some engineer in Burbank or Pasadena or Studio City had determined that I should feel. I'm sitting there watching Jason Statham pummel his way out of a maximum security prison while the world's most incompetent robot masseuse is abusing my chakras. After two hours of this, I stumbled out of the theater having received the message loud and clear that I am *too old for this shit*, an attitude shared by another Gen-X dude behind me, who remarked, "Well, *that* sucked."

The only element of 4DX that I found remotely nifty was the slight ocean breeze that wafted into my personal space when Vin Diesel's Dominic Toretto shows up beachside in Cuba. Otherwise, the experience was exactly the opposite of immersive. The gluteal rumbles and spine-rattling lurches pushed my attention away from the movie, which, incidentally, was a pretty great installment in the franchise. There was Charlize Theron as the baddie, a steely, dread-locked Furiosa by way of Coachela, hell-bent on taking a nuclear sub for a spin. There was Shakespeare-grade dialogue like "I've seen that look in his eyes before. Dominic Toretto has *gone rogue*." And there was the wonder that is Vin Diesel, who acts how a dildo would act if a dildo studied the Stanislavski method. (Bonus: in 4DX, the dildo vibrates.)

Needless to say, the movie left me bruised on a weekend when Father John Misty's new album, full of deprecations of our brave new immersive age, ran in loops in my head and our reality star president rattled his nuclear sabers against an autocrat just as unhinged as he.

Is it enough to let ourselves be entertained, to escape from our news feeds for a precious couple hours to surrender to a chiropractic nightmare? Is this what the arms race of sensation demands? Is our future one in which we'll offer ourselves up to sit compliantly in our seats, our bodies shaken and souls unstirred, or one in which we'll meet strangers from the other sides of the earth in alien landscapes, our identities abstracted into avatars that bear no trace of the races formed in the course of humanity's viral dominion over the globe?

*Fate of the Furious*'s epic centerpiece sequence involves hundreds of self-driving cars that get hacked and driven like a tsunami of vehicular homicide through Manhattan. This is the second movie I've seen recently, the first being *Logan*, which features self-driving vehicles, which we all seem to accept are coming any day now. The act of driving, one of the things we assumed only humans could do, is about to be forfeited to the machines, and therein lies a primal, Luddite anxiety that the movie adroitly exploits. How ironic, then, that a movie that taps our fear of no longer being in control of the wheel has been paired with a next generation cinematic concept that emphatically communicates that the machines have authority over your body, not the other way around.

We find ourselves at a fork in the proverbial road, to belabor a metaphor. One route offers us ever more sophisticated methods to surrender our free will. Another offers human connection in psychedelic playgrounds built in code. Both offer a way to satiate an irrepressible human need for community. I think the real reason why audiences from Hong Kong to Seattle flock to the *The Fast and the Furious* movies, beyond the laughably improbable close calls and military industrial pyrotechnics, is their appeal for us to envision ourselves as belonging to a transracial family spanning languages, continents, and transportation preferences. Just as we yearn to be thrilled, we ache for ways to bridge our differences. I think it is in virtual reality, not cinema, where we're going to cross those bridges.

# WHY YOU LIKE READY PLAYER ONE

A few years ago I noticed that every bookseller I spoke to at Elliott Bay Book Company recommended *Ready Player One*, Ernest Cline's novel about a kid who lives much of his life in a virtual world invented by a 1980s pop culture-obsessed game designer. I read it in a few sittings and enjoyed the hell out of it. It has what I like to call idea density—a steady flow of clever concepts that kept me interested the whole way through. The pacing was tight, the protagonist easy to like, and the cultural references continually pressed my Gen X nostalgia buttons. If you pay attention to the goings on in the VR industry, you're probably aware that *Ready Player One* is given to every new hire in Facebook's Oculus division, and that Stephen Spielberg's adaptation hits theaters in 2018. It's a wildly popular book, deservedly so.

I have been recommended *Ready Player One* by teenage girls and forty-something year-old men. Recently, my girlfriend finished the audio version and it came up during a talk I gave at the Academy of Interactive Entertainment. Published in 2011, the dystopian novel shows no signs of receding gently into the forgotten realms of our culture. But why is the book so popular?

Awhile back Ron Jones of Sky Muse Studios recommended that I read *Hit Makers: The Science of Popularity in the Age of Distraction*, by Derek Thompson. Ron composed music for television for forty some years in Hollywood and knows more than a little about creative hits and misses. I took him seriously when he said that this book distills hit-making down to a science.

One of the main conceits in *Hit Makers* is a principle called MAYA, for "Most Advanced Yet Acceptable." We gravitate to art works that simultaneously reminds us of what we already know and offer some new, novel twist.

A couple weeks ago, in my ongoing effort to impress upon my children just how awesome the nineties were, I showed them *Forrest Gump*, which I hadn't seen since it was in theaters. The movie is a constant bombardment of the MAYA principle, crammed full of cultural touchstones, with a soundtrack that was so extensive it came out on two CDs. Elvis, the Civil Rights movement, Vietnam, the druggy seventies, the eighties' fitness craze, AIDS–it's as though the screenplay was composed by flipping through thirty years of *Time* magazine covers.

*Ready Player One* is like *Forrest Gump* but focused on a single decade and in virtual reality. Ernest Cline and I are about the same age, and his references to Dungeons and Dragons modules, Joust, Rush, and John Hughes movies lit up the parts of my brain that developed in the years when my hair style observed a strict, business-up-front/party-in-the-back policy. One meta layer of the novel is that VR itself is a bit of a late eighties nostalgia button. I remember where I was (Dave Cornelius's history class) when I read that article in *Rolling Stone* about Jaron Lanier and VR.

I read somewhere that Millennials often read *Ready Player One* with a nearby screen open to Wikipedia. Having lived through the era of parachute pants, I didn't need to. So if adults younger than me didn't enjoy the book because of the eighties references, why do so many continue to cite it as one of their favorite books?

Here's where Cline is truly ingenious. In his novel, the MAYA principle works on multiple levels, hitting you in different ways depending on how old you are. If you're a forty-something like me, you enjoy the book for its references to *Ferris Buehler's Day Off* and relive your own adolescence through Wade Watts. But if you're a Millennial, the thing that's familiar is the experience of submerging yourself in digital facsimiles of reality, of screens, of friends who exist exclusively online. The novelty–the thing that's new–is the trove of eighties pop culture.

Cline achieved a remarkable feat, writing a novel built upon a dual MAYA system that adapts to the reader. I'm in awe that he pulled that off, and intrigued to witness how this story further cements itself in popular culture.

# VIRTUAL REALITY VS. THE APOCALYPSE

Wildfires in British Columbia have blanketed Seattle in a haze the color of nicotine on the hottest week of the year. The irony being that it would be a lot hotter if this persistent cloud of particulates weren't hovering over Puget Sound. It's Seafair weekend, which means our annual visit from the Blue Angels, those fighter jets that terrify house pets and trigger PTSD in survivors of war zones. National politics is in full clown car mode, sowing xenophobia, rolling back environmental policy, getting a little too casual for comfort with the idea of deploying nukes. As someone who earns book royalties on an apocalyptic scenario called the Age of Fucked Up Shit, this convergence of troubling signs has me wondering if virtual reality has arrived to save the world.

Does framing VR as an entertainment medium or new investment opportunity sell this technology short? It's when we consider VR's profound therapeutic and empathetic properties, and understand these properties as intrinsic to VR as an art form as well as a therapy, that things become interesting. There are, for example, scientists stationed on a remote base in Hawaii, studying how to live in isolation on Mars by using VR to combat cabin fever. VR promises a level of intimacy with media that we have yet to experience, and this, perhaps, will keep us sane and socially functional in an increasingly hot and hostile world.

It may be that the structure of virtual reality, its capacity to provoke feelings of presence and empathy, has arrived as a solution to one of the greatest problems of our era, the disassociation of our human community. In this country at least, we have never felt as alienated from one another, with common ground being eroded from under our feet by socially mediated, competing versions of reality. As was demonstrated in Spike Lee's *Do the Right Thing*, hot days tend to make us crazier, and every summer feels a little nuttier than the last. Just as we're anxious for a reprieve from the physical heat, so too do we ache for a reprieve from the heat of human conflict.

I hope we're going to engineer our way out of the climate crisis, but fear that even if we do, we'll still endure a wave of suffering the likes of which human civilization has never seen. We have the intellectual capacity to figure our way out of this labyrinth, the collective neocortex of the human race having brought us to a point where we can teach machines to think. Our collective rationality seems to be in a battle with the collective midbrain and limbic systems, those earlier, more foundational emotional and fight-or-flight centers that warn us to fear one another.

And in the midst of this crisis, along comes a medium that invites us to experience the pain of another, to interact as avatars disassociated from our tribal, physiology-based identities. And to do this while closing ourselves off, locking ourselves away in base stations on remote volcanoes, returning to draw, with Tilt Brush, on the walls of our caves.

When I read news of the ups and downs of the VR market and whether or not some new company is going to get another round of funding, I get the feeling that a far more profound process is at play, beyond the fickle horse race of capitalism, more on the level of the agricultural or industrial revolutions. The problems that are coming toward us are gigantic and horrific and, above all, *real*. It remains to be seen whether the solutions to these problems will be up to the task. I have an inkling, a hope, at least, that some of the solutions to these real problems will, strangely, be virtual.

# CULTURE ≈
# WHAT YOU VALUE
# MORE THAN $

I spent much of April, 1998 in a windowless classroom with about a dozen other people learning to master Amazon's elegant, UNIX-based customer database. Most of my fellow reps in training were like me–in their twenties, opinionated about techno music, restless with youthful energy.

Jeff Bezos had famously told the staffing company through which I landed the CS gig, "Send me your freaks." Even among us freaks, Walter stuck out by virtue of his normality. He was in his forties and had recently been laid off from another company. He spoke fluent German and wore khakis, loafers, and Oxford shirts. He never once came across as superior or as a know-it-all, throwing himself into this opportunity to answer phone calls at odd hours for this online bookstore with enthusiasm. When we sat on the floor to eat pizza, Walter grabbed a slice and sat cross-legged with us. By the end of our training month we all looked up to Walter and considered him one of us in every way. Soon, Amazon launched its German store and presumably Walter rocketed up the executive ladder.

Walter taught me an important lesson about the humility required to adapt to a new culture. When you encounter something new as an adult, it's natural to try to understand it by comparing it to what you already know, by making sure everyone knows you're an expert.

Admitting your own ignorance and willingly returning to a more childlike state of receptivity requires guts.

It's hard to communicate how punk rock Amazon seemed in the late nineties, when having no dress code and eating lunch at your desk was considered radical. It really did stand out as an insouciant, maverick company cultivating its own hothouse of relentless innovation. Walter succeeded there because he didn't try to impose the business culture he was familiar with onto an emerging one.

As we create the culture of immersive media, we arrive encumbered with cultures that were incubated in other industries, notably the brogrammer-rich precincts of Silicon Valley. But some of the attitudes that found their most visible expression in VC-fueled tech startups don't jive with what our new, empathetic art forms appear to require from us.

One, the winner-take-all clamor for market dominance, the "we own this space" mentality, is anathema to a medium that promotes empathy. Considering that young people are increasingly skeptical of capitalism, one might argue that the economics of immersive media are primed to return to the more communitarian spirit that animated the Homebrew Computer Club where Jobs and Wozniak got their start. Startups that understand how to embrace the zeitgeist will operate more collaboratively than those who blindly accept the corporate Darwinism that animates Facebook, Twitter, and Uber.

Startup culture, in Seattle at least, is slowly evolving to become more female. Traits that we've been conditioned to consider "feminine," like cooperation, nurturing, and compromise, are advantageous in a medium that's by nature social and empathetic. The sexism that's rampant in game studios and other tech companies is increasingly proving to be not just morally wrong, but malignantly cancerous to the bottom line. We have a long way to go.

If we can point to culture as the reason why companies like Uber stumble, then it follows that culture can be the reason why certain other companies succeed. It bears asking: what kind of business culture do we want to cultivate in the immersive media industry?

Culture is stubbornly subjective and difficult to quantify. It's ultimately an expression of values. What are values for, exactly? Values are what guide us through the gray areas of our lives and help us make decisions in ambiguous circumstances. Values are adaptive, and adaptiveness is scary because it appears to be noncommital. Businesses traditionally tend to favor predictability over squishiness. I worked for a number of dotcom companies that gave lip service to Tom Peters-ish notions of thriving on chaos but at the end of the day were resolutely biased toward what worked in the past. To take risks, you must be guided by faith in your values.

Metrics-minded people who attempt to solve the problem of culture can fall into the trap of over-quantification. Questions of how to make a workplace more inclusive become band-aid diversity initiatives designed to flatter white progressives more than actually build on-ramps from underrepresented communities into an industry. Corporate retreats and team-building exercises come across as laughably disingenuous, and everybody secretly knows that "no idea is a bad idea" is a lie, that expressing a weird idea makes the higher-ups question whether you're a "team player."

Cultures can be built upon the rickety scaffolding of platitudes. I've worked for bosses who trumpet their "open door policy" then hide in their offices avoiding conversations with their subordinates. Organizations that call themselves flat are often governed by unspoken hierarchies that favor the extroverts. The most horrible meetings I've ever endured have always been the ones in which everyone's opinion supposedly carries equal weight, which means that the most sociopathic loudmouth in the group gets to set the agenda and ruin everyone's lunch.

It takes more than perks and team-building retreats to create a culture. I think one way to assess a company's culture is to ask what it values more than money. Another way to put this: what are you willing to go out of business for? Here's a thought experiment. Imagine yourself on your death bed, thinking about the company you founded that went out of business. If by going belly-up, you honored your values, then you won. Every company is going to go out of business one day, just as every human being will one day be a rotting corpse. Seriously: remind yourself that you will soon be dead and, chances are, few people, if any, will give a fuck. Then embrace this knowledge and use it as fuel to do something awesome for other people. The most we can hope for is that our name will be spoken throughout history and associated with values that help make life easier for the people yet to be born.

Personally, as an endup founder, what I value more than money is an easy question to answer. I value art more than money, at times to the considerable chagrin of my parents. My whole adult life seems to have been an effort to fool people into giving me money so that I can get away with my creative expression. I'm intent on gaming the system well enough so that I can write books and now create VR content. Don't get me wrong, I like money, too. But I see money as the means to get to do the work I truly want to do, rather than the reason for doing the work.

The paradox of this attitude seems to be that my proudest achievements have always been the result of prioritizing art over money. In other words, if you're not pursuing money in order to fill some ravenous, emotional hole in your life, you'll eventually be rewarded for pursuing what you value most. I recognize that there's a leap of faith in that statement.

In practice, this attitude is complicated and not guaranteed to pan out. But I have to believe it's better to spend a life being rewarded by the pleasure of the work itself rather than pursuing something merely tolerable in order to acquire status symbols. I've noticed that the most miserable people in the world seem to share the trait that they never made the leap to pursue what they were most passionate about, and chose to bargain away their passions for short-term security and status.

The question that every VR startup founder must face when contemplating what kind of culture they wish to cultivate is: what do I value more than making a buck and getting bloggers to think I'm cool? If the answer is building a community, creating opportunities for those who've been marginalized, or bringing beauty into the world, then I bet we'll begin to discover properties of this new medium beyond what we can now imagine, and the reward will be far more magnificent as a result.

# WHAT I'VE LEARNED FROM PLAYING VIDEO GAMES

The more I hang out with video game designers, this creative, financially stable tribe, the more self-conscious I become about my own lack of gaming knowledge. I realize I come across as pretty uninformed, with my points of reference stuck resolutely in the mid nineties–*Myst*, *The Sims*, *Civilization*.[1] I knew that if I was going to work in virtual reality, I at least needed to be conversant in some of the gaming conventions that were being ported over from the world of consoles and RPGs. So I threw myself at the mercy of my social network and asked if anyone would take pity on a poor book nerd such as myself, lend me their console, and suggest a couple games. I expressed particular interest in open world, explorable sandbox games.

Members of the local VR community were characteristically generous, and pretty soon I found myself with an XBox One, two XBox 360s, and lots of games. I felt a bit like Christopher Plummer's character in the Mike Mills film *Beginners*, discovering house music at age seventy. It's both humbling and exciting to explore a whole method of entertainment that you've otherwise ignored.

I was intimidated by the complexity of console games. The vaguely

---

1    I've been fortunate to get to know Robin Miller, an all-around cool fellow and co-creator of the blockbuster *Myst*. When I met him for coffee the first time, I confessed that the level of obsession I'd experienced while playing *Myst* had made me swear off video games for twenty years. "You're welcome?" he said.

amorphous controller with its multiple buttons and two joysticks always looked complicated; I couldn't imagine ever mastering it. As I loaded up the first of what would prove to be a series of deep gaming experiences, *Red Dead Redemption*, I felt as though the reprogramming of my brain had begun.

I played *Red Dead* for a couple months, ignoring the narrative challenges, choosing to simply wander the dusty trails and snowy woods of this version of the mythic, gun-slinging West. The vastness and detail of this geographical facsimile of American terrain—mountains, prairies, coastal towns, desert—was itself entertaining enough for me. In *Red Dead* you play a grizzled cowboy named John Marston, and in between the occasional gunfight with outlaws and random cougar attacks there are moments in which you can simply take in a spectacularly rendered vista. I found myself placing Marston at the edges of cliffs or on train trestles then using the POV joystick to swoop around him, as if I was directing the movements of a camera in a movie.

Mostly I had Marston hunting and collecting pelts, harvesting wildflowers, and seeking out unexplored corners of the map. The game tapped the same story-making impulses I had enjoyed as a kid playing in the woods. One night, I had Marston wander into the city of Blackwater, and I decided that he was going to seek revenge on a man who had assaulted his daughter. This was entirely my own narrative and didn't have anything to do with how the designers built the game. I trailed a random villager, a non-player character governed by artificial intelligence, some shop keeper. He behaved as if he was unaware I had shown up in town to exact my revenge, and this added a new layer of dramatic irony to the game. I knew that as soon as Marston put a bullet in this guy's skull that the other NPCs would spring into action and there'd be a massive vigilante shoot out, so I carefully plotted the assassination.

This is to say, I made my little video game character wander through a 3D town and created meaning by inventing a story in my head. It occurred to me that video game narratives can replicate the experience of writing stories as much as passively receiving them. I could make my character do any number of things, but the reasons, the connective tissue of causality, were largely within my purview to provide.

Spoiler alert, I managed to murder that bad hombre and high-tail it out of town. Then I descended into Tarantino mode. I wondered what would happen if I walked into the Sheriff's station with hundreds of rounds of ammo, shot everybody, then hung out waiting for the posse to show up. So I performed this homicidal psychopath routine for a while, filled the station with corpses, got killed, and was resurrected anew with my moral slate wiped clean, free to gather wildflowers on the high plains again.

I grew up in an era when an alarming number of actual, voting-age adults believed that a metaphysical being from another dimension was personally interfering in the audio engineering procedures of heavy metal bands, encoding "backmasked" pro-suicide messages on the vinyl grooves of LPs. I recall the hysteria of the Parent's Music Resource Center, whose warning label efforts proved to be an effective marketing tool that pointed me in the direction of the music I most wanted to hear. I also remember watching a made-for-TV movie starring Tom Hanks about how playing Dungeons and Dragons was inspiring real-life ritual homicide. Hysteria over violent entertainment is nothing new.

You just can't make a causal connection between real-life violence and video game violence any more than you can make such a claim about Cormac McCarthy's *Blood Meridian*, a novel far grislier than what Rock Star Games crafted in *Red Dead*. Symbolic violence is just the surface level of what these games seem to really be about—overcoming challenges to acquire resources, and in the process advancing one's capacity to overcome challenges and acquire resources. You kill things, take things from the things you've killed, and that makes it easier to the kill and take from the next thing.

After *Red Dead Redemption* I spent several weeks in the post-apocalyptic Boston of *Fallout 4*. While *Red Dead* has you occupying the predetermined role of a particular cowboy, you can customize your *Fallout 4* character. I chose to make mine a woman I named Anne Specularian, the recently thawed-out resident of an underground bunker who finds the retro-futuristic landscape overrun by mutant naked mole rats, ghouls, marauders, and an economy based on bottle caps. There's a cheeky, winking edge to *Fallout 4*'s Tomorrowland version of the post-apocalypse, and I was soon absorbed in the splendid grossness of it all. The mutants are hulking green brutes, the mutated bugs ooze toxic juices, and piles of garbage and urban decay make the whole affair vaguely depressing even when the combat is thrilling.

One of the storylines of *Fallout 4* allows you to join an organization called the Brotherhood of Steel, a heavily armored paramilitary force whose command center is a dirigible that hovers as a rare outpost of order and decorum above the bombed-out Boston airport. The violence that you deliver to various undead and mutated residents of Beantown comes to feel justified by the comforts of the civilization that has agreed to keep you well armed. You can join up with other bands of resistance, too, and help construct shelters and other resources for ragtag communities of hapless farmers and settlers. Over time the various baddies that can be dispatched with a baroque assortment of armaments appear to exist to refill your adolescent power fantasy meter. Thank goodness there's another raider around the corner, giving you another opportunity to diversify your homicide skills.

I encountered an unexpected moment of tenderness in *Fallout 4*, when I had an opportunity to exchange my companion, a super mutant named Strong, with a cocky douchebag of a mercenary. When the game asked if I wanted to send Strong away, the vaguely gelatinous-looking NPC delivered a few lines of sad dialogue asking why I was rejecting him. And it really did get to me, to the degree that I changed my mind and sent the mercenary packing instead.

I paused the game and reflected on what had just happened. I'd had an emotional experience as poignant as any I've had watching a movie or reading a book. That this experience involved a brutish monster inclined to crush radioactive bugs with a sledgehammer seemed beside the point. More salient was the deftly crafted illusion, bolstered by the martial rhythms and orchestral swells of the soundtrack, that one of my decisions could emotionally impact another sentient being. I considered how thoroughly I'd been seduced by this illusion.

I've spent a lot of my life working out how to manipulate peoples' emotions with words designed to be read in a certain sequence. Narrative, to me, has always been overlaid on the passage of time, and stories have to proceed in a very specific order in order to generate meaning. My experience with open world games has introduced new and exciting possibilities of narrative choice into the concept of story.

The orientation between storyteller and audience changes when the audience is given agency—it's not simply that there are more paths available for the audience to follow, it's that by providing multiple paths, the creator is ceding some of their own responsibility of cultivating meaning, handing the audience some of the tools of the writer.

Having reached points of exhaustion with *Fallout 4* and *Red Dead Redemption* I ventured into a game that many of my gamer friends claim to never have fully exhausted, *Skyrim*. It was as vivid and wonder-filled as they'd claimed, and I spent a good month probing its dungeons, defeating dragons, and enduring its inscrutable dialogue, the kind of Tolkienesque huffing and puffing that the writer Tom Bissell calls "lobotomized Shakespeare." This isn't to discount how fantastically creative the game is in so many ways, but like the Wild West of *Red Dead* and the bombed out world of *Fallout 4*, *Skyrim*'s aesthetic M.O. appears to be trope-based.

I came down with a gnarly flu while playing *Skyrim*, and perhaps it was the illness overlapping with the kill-and-plunder mechanics of the game, but the whole experience started to make me feel sick. I started getting grossed out by the process of slaying another frost troll, looting yet another treasure chest. It seemed that my capacity for this kind of entertainment might have run its course, at least for the time being.

Over my several months of immersion into video games, I'd been wowed, moved, dazzled, and impressed. But did these experiences make me a better person? Had I learned anything while playing video games that I could apply to my day to day interactions with flesh and blood human beings?

Just a few days after the grand guignol of Trump's inaugural, I watched Martin Scorcese's *Silence*, the story of two 17th century Jesuit priests attempting to minister to Japanese Christians in a society hostile to their faith. The film asks us to consider what we're willing to make others endure for our beliefs, and has much to say about empathy, cultural dislocation, and power. *Silence* made me reflect on how to become a better human being and rattled me profoundly. Later that evening, I played a few minutes of *Skyrim* and the activity seemed kind of a bullshit waste of time.

We yearn to get outside our own heads and into the heads of

others. We reflexively gravitate toward eavesdropping on the broken hearts and scheming minds of people who are unlike ourselves. It's incredible how much we indulge our fantastic worlds to obliterate our loneliness in the presence of others. I suppose this is ultimately what compels us, as children, to play.

It's clear to me that video games have evolved into a powerful art form, and I know I've just scratched the surface of what's possible, in terms of emotive connectivity, with the medium. Games are more diverse and sophisticated than most people give them credit for, but it is interesting to note how many of them still seem morally constrained by the structure of kill/acquire/level up.

I sense that gaming itself is in the process of leveling up as it sets foot in the realm of VR, where it will aim not just to entertain but to challenge, to rattle, to question. We call our entertainments escapism but the best of them force us to confronting the very things we would most prefer to escape.

# IMMERSIVE MEDIA
# IN RURAL AMERICA

On the dark and rainy morning after the shocking 2016 presidential election, I found myself in the control room of Sky Muse Studios, a recording facility in the woods outside Stanwood, a couple miles from the house I grew up in. My childhood home was a half mile north of the Skagit-Snohomish county line, within earshot of Interstate 5, on seven acres of fields, forest, and ponds in what often felt like the middle of nowhere. This new studio is practically on my grade school bus route.

It's hard to overstate just how isolated rural places like my home felt before the Internet. The world was awash in entertainment and culture that originated from New York and Los Angeles, delivered in the form of TV signals captured by an antenna and weekly magazines that showed up in our mailbox. Most of the world didn't seem to know or care that Seattle, much less Skagit County, even existed. I knew, growing up, that I wanted to contribute to American culture in some way, most likely by writing novels. I always believed that there was no reason why someone from a small place couldn't grow up to influence the world. By leaving that small place for the big city. But I never imagined that technological revolutions would be happen right where I used to ride my bike and climb apples trees.

Sky Muse Studios is the brainchild of Ron Jones, a composer

with a mind-boggling list of credits to his name. If you've turned on a television in the past forty years, you've heard Ron's music. Ron was the musical mind behind *Family Guy*, *Star Trek the Next Generation*, *The Smurfs*, *Scooby Doo*, and myriad themes and motifs that have gotten stuck in heads and hummed on playgrounds. When he was still in his twenties, Ron could toggle between NBC, ABC, and CBS on his television set and hear music he had composed on each network. After his career in Hollywood, Ron decided to return to the Pacific Northwest of his childhood (he grew up in Bellevue), and ended up on over 20 acres just outside Stanwood, a friendly community known for its Scandinavian heritage and vegetable processing plants.

I visited Sky Muse with audio mastering pro Steve Turnidge and Christopher Hegstrom, a co-founder of the organization Audio VR. Both big-time audio tech geeks. And while most of the audio tech jargon soared right over my head, I could tell from the puddles of drool forming under these guys that Sky Muse is a sophisticated operation, with space and gear to record everything from orchestras to singer-songwriters, all arrayed on a property surrounded by trails, woods, and fields. There's even a guest house nestled among alders where a band can chill out and/or record.

Over breakfast at Wayne's Cafe in downtown Stanwood, Ron shared his story and vision for Sky Muse and I marveled at how a Hollywood-quality recording studio ended up so close to my childhood stomping grounds. For almost twenty years I've lived in Seattle, and I identify myself as an urbanite, with the progressive values that go along with that identity. But I also understand rural Washington state as someone who worked in the tulip fields for a summer and spent my childhood among barns, tractors, and livestock.

We've heard that rural white voters propelled Trump to the presidency. And that one reason why these voters went so decisively for Trump was because jobs in rural America have dried up, particularly jobs in manufacturing. Trump perpetuated the lie that the reason these jobs disappeared was because of immigrants and outsourcing. The truth is, manufacturing has been transformed by automation, and industries evolve and die off all the time. Maybe it's just harder to stoke a racist bias against robots.

I can understand how people living in rural America don't feel included in the culture they see streaming from their devices. This doesn't let anyone off the hook for bigotry, xenophobia, misogyny, or the laundry list of outrageous positions Trump and his voters championed. But the divisions between rural and urban America are so ingrained that they often don't seem worth questioning. One such division is the idea that big cities produce culture, and rural areas only receive it. If you grew up a kid like me who was obsessed with music, film, books, and the media, then leaving rural America for a city was a foregone conclusion; when I graduated from high school I couldn't get out of Mount Vernon fast enough. Rural America is good at pushing its creative people away into cities where they can find one another.

Over time this division has solidified into an assumption in rural America that you have to move away if you want to contribute to any kind of cultural economy. This assumption is deep and persistent and hasn't necessarily kept pace with the connectivity that the Internet enables. Someone playing a guitar in a cow pasture can upload their video to YouTube just as easily as someone playing a guitar in a club in a city. The Internet has helped rural America get far more sophisticated and globally minded than city dwellers give them credit for. And yet there's a persistent assumption that the apparatuses that produce culture exist exclusively in the metropolis. Part of this, I recognize, has to do with the heightened level of competitive friction you can only find in a city.

Later, I returned to Sky Muse Studios to watch a documentary called *Fight for Space* about NASA's somewhat checkered history. The feature included archival footage of various missions and launches interspersed with commentary from astronauts and public scientists like Neil Degrasse Tyson and Bill Nye. All the music and sound for the film was produced at Sky Muse under Ron's guidance. Before the screening, Ron spoke a bit about how the studio was conceived two years ago and how he assembled a team of young engineers and musicians from area community colleges. His audience of twenty or so guests sat in fold-out chairs on risers in the room where the film's music was recorded.

It was easy to connect the documentary's appeals to dream big and Jones's vision to establish a Hollywood-quality recording facility in the woods. Sky Muse Studios is a bit of a moon shot itself, fueled by the belief that movies, TV shows, and virtual experiences can be produced in Snohomish County just as well as anywhere.

Ron spoke with pride about the musicians and engineers who'd worked on the film. He'd tried to no avail to lure pro talent from LA and Seattle, so he turned to Digipen, Cornish, and other schools in the region to recruit his team. They produced a soundtrack that's grand and period-appropriate, with passages of acid rock and awe-struck orchestral swells. In a follow-up email, Ron said, "Sky Muse builds people more than audio or video tracks."

I think the lesson here is that you don't create something great by looking over your shoulder at some other place that's known to produce greatness. You survey your own landscape, take stock of the resources at hand, and invest your time and faith in dedicated people who share your vision.

All the world's centers of culture started as humble places populated with just enough dreamers. Watch an old Buster Keaton film. It's fascinating to see silent images of the Los Angeles of the early twentieth century, all dusty roads and lemon groves. Ashland used to be just some town in Oregon before it became known for its Shakespeare festival. There's no reason that Stanwood, Washington can't thrive, providing the technology and high production values of Hollywood amid grazing horses and salmon-packed creeks trickling through forests to Puget Sound. It's a place where immersive audio technology under development in Seattle can be tested under the trees that give the Evergreen State its nickname. It would behoove Stanwood, Mount Vernon, and other nearby towns to consider how to commit to the success of places like Sky Muse Studios. A big first step is embracing the idea that places like Stanwood have what it takes to become important contributors to the immersive media industry, and that the people who can do it are close at hand.

# INVEST BROAD AND SHALLOW, NOT NARROW AND DEEP

At a recent Seattle's VR/AR meetup at Pluto VR's Ballard office, it struck me how much the community has grown in the short time I've been a part of it. I look forward to seeing familiar faces at these meetups and enjoy catching up with people like Andrew Mitrak, Eva Hoerth, Jordan Kellog, Trond Nilsen, Bridget Swirksi, and many others. The same core of pioneers continues to show up while new folks are being welcomed into the fold. I talked to Joshua Jonas about his successful trip to China and with Dave Huber about his audio collaborations in Amsterdam and Berlin. There were new demos, reports of breakthroughs, and sneak peeks at mind-blowing experiences to come. All in all, everyone seems to be making progress.

But I also detected an air of sober uncertainty. The meetup happened on the night that Altspace VR announced its plan to shut down. Reaction to that news was rueful but not shocked. Tellingly, no one deemed the demise of much-admired Altspace a failure, but rather a necessary step in the evolution of social VR.

I'm fascinated by the process by which we adapt knowledge gained in one area to a new area. For instance, as we try to figure out storytelling in VR, we refer to lessons we learned in older media like movies, games, and literature. The challenge is to listen to the new medium and be receptive to the ways in which it wants to diverge from the structural properties of previous media.

While talking to Joshua Jonas, who has been neck deep in the fundraising process as Inverse Studios's CEO, it occurred to me that the investment community is also applying what it learned from previous industries. It seems that VCs are obsessed with finding the one company that will break out and deliver VR's killer app, driving higher adoption. If your primary goal is to make money and you're investing in a new industry, then it's inevitable that a gambling mentality can take hold. When a company like Altspace VR goes under, it's widely interpreted as a sign that somebody didn't bet on the right horse.

Instead of thinking about the VR industry as a collection of companies that are each competing to be the Twitter or Google of VR, we can—we must—think much more holistically. There is a difference between investing in a company and investing in an entire creative economy. The trouble is, the channels of investment are securely established toward individual companies, not ecosystems, sustained by the myth of the charismatic boy-creator. As I looked around the room at the meetup, I wondered what would happen if every person in attendance was given $10,000 to work on something related to VR. My suspicion is that the 100 or so people would form valuable partnerships and alliances and pool their resources into creating experiences that any single company wouldn't be capable of creating.

The VR creators I know who are heavily engaged in the investment process all seem to be playing somebody else's game, a game designed by Silicon Valley's startup culture to maximize the possibility of a breakout company. The problem isn't that VR companies like Altspace aren't succeeding in this model—the problem is the model.

Ideas have a better chance of taking root and growing the more surface area they're allowed to cover. Ideas thrive when shared. Creative communities like Seattle's VR scene operate as idea distribution platforms that also encourage ideas to combine, mutate, and give birth to other ideas. If you're intent on getting a return on your investment, you're more inclined to storehouse and hide ideas. This seems to be the primary tension at play in the VR industry right now. On one side are communities of independent creators motivated to spread ideas around, and on the other are speculators who want to make money by hoarding and claiming as many good ideas as possible.

A truly visionary investor would create a community slush fund that would give out many modest grants to, for instance, allow hackathon teams to invest in better gear. My hunch is that in the long run, giving out a little money to a lot of people is better than giving out a lot of money to a few people. The point shouldn't be to go all-in with a group of ten developers working in secret in a garage. It should be to help as many VR creators as possible pay the rent for the next three months, freeing them up to collaborate and discover the unexpected ideas that arise when they have more territory in which to grow.

There are examples of these sorts of small investments paying off in a big way. Stockholm, Sweden, today is one of the biggest music production centers in part because in the 1970s, the government expanded music education and gave micro-grants that helped musicians buy instruments. This alleviated some of the economic pressure at the grassroots level, allowing for artists to develop without having to worry so much about paying the bills. Imagine what would happen in Seattle if the 100 or so people at last week's meetup were given Vives.

Status thinking prevents us from seeing how effective a strategy of wide and shallow investment could be. To horribly mangle three metaphors, we're all hoping to lasso the big enchilada, when we'll rise together if our personal ambitions take a back seat to our community ambitions. Imagine the power in not just striving for your own company's success, but being equally invested in the success of three or four other companies. (Again, think of Amazon's Marketplace.) This is what I mean when I talk about competing like artists instead of athletes. When you compete like an artist, you rely on your competition for inspiration. And in Seattle's VR community, there is plenty of inspiration to go around.

# MICROSOFT'S REFRESHING NEW PERSONALITY

If Microsoft were a person, I'd offer that for most of the company's history it has been an affable soccer dad, his hair swept back in a ponytail, sporting a goatee, glasses, and a cell phone clipped to his belt, driving around a Costco parking lot in a PT Cruiser blasting a bootleg of a band that plays rock operas in time signatures other than 4/4. A guy you'd have an entirely pleasant conversation with at a barbecue about spooky action at a distance, but who could quickly corner you and talk your ear off about cryptocurrency if you weren't careful. Easy to like, if a bit socially awkward, with self-confidence bordering on smugness. Spends his money on expensive audio equipment and has pictures from a family trip to Italy on his desk. Was way into fist bumps before any other white person you knew. Uses the word "webinar" with no sign of lexical embarrassment. Wears Seahawks jerseys on game days.

That's the Microsoft I knew when I worked there for a blink of an eye in 2001, my two-month tenure as a contractor editing documentation related to the Xbox bisected by 9/11. During those two months I imagined myself as a droplet in a vast ocean. I worked out of an office in Issaquah and marveled at the company's size, with its own fleet of shuttles, massive food service facilities, and picnics featuring the musical stylings[1] of Cheap Trick. When I got a full-time job with a Vulcan-backed startup, I left Microsoft with their mantra, "eat your own dog food" ringing in my head.

"Eat your own dog food" is a concept that came to define Microsoft for me in that era. It means using your own products even if there's a product on the market that's superior to yours. This meant that instead of using the web-authoring software Dreamweaver at Microsoft, I used Microsoft's web authoring tool, whose name I can't remember. While parked at my desk in a windowless office in the misty Cascade foothills, I occasionally muttered "Microsoft Microsoft" to the tune of the song in Being John Malkovich, in which the titular actor finds himself in a hellish, self-referential dreamscape where everyone is a clone of himself, including a lounge singer who can only sing his name.

That's to say, Microsoft had become massively self-referential. Every business uses Microsoft products; at Microsoft they use Microsoft products to make more Microsoft products so that more people use other Microsoft products. I came to think that the company's Clinton-era antitrust scrimmages weren't so much the result of venality on the part of Bill Gates, but of a holistic obsession to ensure that every piece of Microsoft software worked with every other piece of Microsoft software. As a result, the company developed an insularity that blinded it to opportunities that companies like Google, Apple, and Amazon were set to exploit.

I still thought of Microsoft in these terms when I signed on to participate in the Hololens Hackathon in the spring of 2016. I set up my MacBook Pro at a table at Fremont Studios and immediately felt conspicuous, like I'd get busted for using an Apple product. One of the Microsoft employees who was on hand to assist with the event disabused me of this notion. She explained that nowadays you could walk across Microsoft's campus and see people using Android and Apple products all the time. There was a new openness to Microsoft's culture, a platform agnosticism that represented a break from the eat-your-own-dog food era I had encountered in 2001.

Over the course of that weekend, I got to experience the Hololens, an astonishing augmented reality device that seemed like science

---

1    Isn't it funny how the word "stylings" makes you instantly suspect it isn't any good? Same holds true for "interpretive" dance.

fiction. And my impression of Microsoft started to change. The dad wearing a Bluetooth headset as an accessory started to morph into a leaner, less Caucasian, younger sort of character. Still a guy, sure, but maybe a guy who does yoga and reads poetry.

Now I understand that this cultural shift came about largely as a result of Satya Nadella taking over as CEO, an experience he recounts in his new book, *Hit Refresh*. Nadella sees his role as steering the company's culture toward more empathetic engagement with its customers, and identifies three areas in which Microsoft will grow: mixed reality, quantum computing, and artificial intelligence. It's in the confluence of these technologies that the CEO sees the most promise.

I appreciated Nadella's literary references, though I did take issue with his notion of an AI that could help him write like Faulkner. I'd argue that the only way for Nadella to write like Faulkner would be to experience Faulkner's life, from the racism he observed in the American South to the way his heart was stirred by a hymn. (That said, I have some ideas about how to teach an AI to write a novel.)

Having gotten a taste of Microsoft's version of mixed reality, and having observed the company's newfound appetite for a wide assortment of other peoples' dog food, I'm suddenly interested in Microsoft products again. I recently took my kids to the Microsoft store at Bell Square Mall in Bellevue to try the Microsoft Surface Studio. It's the first example of Microsoft operating at Apple's level in terms of design.

My daughter picked up the stylus, twisted the Surface Dial, and started drawing on this elegant machine. An Indian gentleman, seemed like a MS engineer, stopped by, showed her some cool feature, grinned and said, "Enjoy!" before walking away. That moment seemed to encapsulate much of the warm spirit I had been absorbing from Nadella's heartfelt book.

Then the Surface Studio next to the one my daughter was using crashed for no discernible reason. Ah, good old Microsoft. We missed you.

# TWENTY YEARS
# IN SEATTLE
# WITH AMAZON

I moved to Seattle in September, 1997, to an apartment with a view of the Acacia Cemetery on Lake City Way. For entertainment, I'd often stand at the window with a cup of coffee, watching hunched people sobbing over graves. I'd given up a cool job at a bookstore to follow my girlfriend as she pursed a Naturopathic medical degree at Bastyr. I was in graduate school myself, enrolled in a low-residency Master of Fine Arts program at Bennington College. The low-residency format meant I could live anywhere, and in the back of my mind I had long known I would eventually end up in Seattle anyway. I remember those first few months in Seattle as particularly lonely.

After a short stint working for a sad little poetry press run out of a Green Lake garage, I got a job answering phones and email for Amazon.com in the spring of 1998. That was a season of 90 straight days of rain, and we're talking raindrops the size of Tic Tacs. It was all the same to me, as I spent much of that season in a windowless call center downtown. If memory serves, the Customer Service department at Amazon was all of 200 people. Within a couple months of my start date, we learned that the company was going to branch out into selling CDs and VHS tapes. This e-commerce startup was really going places!

I remember certain books I read then. Thomas Hardy's *Jude the Obscure*, Thomas Bernhard's *The Lime Works*, Amy Hempel's *The*

*Dog of the Marriage*, Virgil's *Aeneid*, and Don DeLillo's *Underworld*. I remember the late nineties mini-explosion of American auteur cinema—*Buffalo 66*, *Being John Malkovich*, *Magnolia*. Late nineties, post-Cobain Seattle was awash in either electronic music or throwback bands that your grandparents could dig. Combustible Edison, Fatboy Slim, Underworld, Air, Big Bad Voodoo Daddy. Radiohead's *OK Computer* and *The Soft Bulletin* by Flaming Lips came along to flush the taste of Smashmouth out of our mouths. Overnight, it seemed, everybody had started dipping bread in olive oil and balsamic vinegar, hosting cocktail parties, and believing in the prognostications of *Wired* magazine. It was fashionable, in those days, for paradigms to *shift*.

Hovering at the end of the 20th century, anxieties percolated over the global catastrophe to come when computers would blank out on what year it was. In November, 1999, Seattle erupted in riots as the World Trade Organization convened downtown at the Convention Center, an event I captured on Super-8 video tape. News of the world still came primarily through TV and newspapers dropped every morning on the welcome mat. President Clinton ejaculated onto a Gap dress. The Chicago Bulls dominated. Napster showed up to offer a treasure trove of bootlegs. Ostensible adults bought black PVC pants at Hot Topic.

I moved to First Hill, to a new apartment with a view of Amazon's new headquarters, the Pac Med building perched on the north edge of Beacon Hill. Some days I would have meetings in that building's 8th floor conference room, from which I could see my apartment. That conference room also provides one of the least-appreciated views of downtown Seattle, from the southeast, with I-5 and Interstate 90, two blood streams spanning continents, converging at the King Dome.

Ah, the Dome. I watched Ken Griffey Jr. smash homers from the nosebleeds in that poured concrete clamshell of a stadium. And then I watched it collapse from an office in the nearby Columbia Tower. When the Dome imploded, a great gray cloud flowed through the streets of Pioneer Square. I watched one pedestrian running like hell to escape the plumes of obliterated concrete, and afterward people writing goodbye messages to the stadium in the dust that accumulated on parked cars.

The other day I woke up to find ash from the wildfires to the east covering the sills of windows I'd left open overnight. A rush of associations... Faint echoes of the eruption of Mount Saint Helens, which I'd heard in my family's kitchen. Anxiety about the Pacific Northwest being behind schedule for a major earthquake. Worries about driving on the rickety viaduct. The last time we got hit by a major seismic event, the Nisqually quake of 2001, I wasn't in Seattle but in upstate New York at an artist colony. I assumed the jackass playwright who broke the news to me was pulling my leg. I found a bar broadcasting footage of the quake and tried to contact everyone I knew in Seattle with a cell phone that couldn't text, take pictures, or play music.

I left Amazon in 2000 to ride a dotcom roller coaster through a series of Internet startups. I initially second guessed my decision to jump ship, then realized I'd made the right choice when, a few months later, the company laid off the entire Customer Service department. The severance packages my friends and former colleagues received seemed to me a sign of executive guilt.

I returned to Amazon in 2004, initially as a contractor writing and editing standardized customer email responses, or "blurbs" for exceedingly polite call center employees in India. Soon I secured an especially sexy job, Editor on the Media Merchandising team. Weirdly, I ended up sitting at a desk next to an editor, Ben Reese, who I'd sat next to four years earlier when we both worked in CS. My responsibilities included crafting email campaigns and on-site messaging and occasionally flying to Hollywood to interview movie stars. I asked Jane Fonda if she approved of stealing office supplies, and Gene Simmons if he'd ever considered donating his vast collection of Polaroids of sexual conquests to the Smithsonian. When Steve Jobs navigated to the DVD store during the iPhone launch event, some copy I had written about a DVD sale appeared on the screen over his shoulder.

A decade ago, Amazon occupied Pac Med, the US buildings in the International District, and parts of Columbia Tower. As was the case the first time I worked at Amazon, I moved offices often. The company was growing for sure, but it had yet to colonize an entire neighborhood. That would happen after I left the second time, much more unceremoniously, in 2007.

My feelings about Amazon over the years have been so complicated that I didn't order anything from them for a decade. Now I'm a Prime member and have an Amazon Echo. Robert Frost's grave stone in Bennington, Vermont bear the words "I had a lover's quarrel with the world." I feel much the same about the company that Jeff Bezos founded. For years I fell more on the quarreling side of that equation, especially as I plied my trade in a publishing industry that Amazon was sticking its forks into. Now I find myself, twenty years on, with news that the company is searching for a second city to colonize, in admiring awe at what this scrappy challenger to Barnes and Noble has pulled off.

One thing that has become apparent to me in reading the reaction and analysis of Amazon's HQ2 announcement is that most of Seattle still has no fucking clue what Amazon is about or how they think or operate. Amazon wins games by changing the way the games are played then being the only ones who understand the new rules. Meanwhile, the company's critics fail by simply sticking to the old rules and insisting that those rules are somehow sacrosanct. I watched this painful process happen to a publishing industry that published four of my books. Seattle's civic psyche still bears the scars of Boeing's corporate departure, and Amazon's news seems to have triggered memories of that break-up. But HQ2 is something entirely new.

Open a tab and go to Amazon. Now, imagine that two other people are also opening tabs and going to Amazon. Each of you is looking at a different store, with messaging based on your individual purchase and browse history. Or maybe some department at Amazon is conducting a test on whether more customers click through messaging related to Prime Music or the Echo Dot. Every single thing you see on Amazon is tracked and analyzed; your interactions with the site producing a rivulet of metrics that join a surging river of aggregate data. Amazon is constitutionally designed to obsessively test itself and try new things. And the company is constantly pitting one idea against another, in a process called A/B testing.

Up till now, the concept of A/B testing has been expressed primarily through the website itself, and through its hyper-competitive corporate culture. When the announcement came that Amazon would open not a satellite office, but a doppelganger HQ, my first thought was holy shit, they're going to A/B test the entire company.

Some of Seattle's civic leaders are sniveling that Amazon doesn't love us anymore and a shudder just went through our white hot real estate market. These folks are, again, under the mistaken impression that Amazon thinks like Boeing in 2001. Companies embrace a city as their sole HQ or choose to uproot and move. That's the way the game is supposed to be played. Amazon is doing neither. Opening a new headquarters that will mirror what's still rising in South Lake Union gives the company an opportunity to build redundancies and compete against itself to arrive at the most innovative ideas. Why? Because they can afford to. HQ2 is also a hedge against the apocalypse, should the tsunami or one of Kim Jong-Un's warheads decimate Puget Sound. The company is thinking far more expansively than most people are giving it credit for, reinventing what it means to have a corporate headquarters.

I attended maybe half a dozen meetings with Jeff Bezos during my two stints at Amazon. The most memorable one was in the winter of 1999, in a conference room in the cold TRA building downtown. I sat beside him among a couple dozen other Customer Service Reps packed into the room. He still had hair in those days and we always knew when he was in the building because his distinctive, braying laugh would ring out among the cubicles. And he was laughing his ass off in those days. How could he not be? The company's meteoric rise often seemed ridiculous, especially to those of us inside of it, and probably to him most of all. But what he told us reps that morning was sobering: wake up every day terrified that everything could fail and never get lulled into complacency. At the end of the meeting, we all rushed back to our desks to jump on calls as quickly as we could, eager to lay the bricks of empire.

# FOUR ESSAYS ON ARTIFICIAL INTELLIGENCE

I. AI-phobia

Who's afraid of artificial intelligence? Plenty of people, it turns out. What is everyone afraid of? Simply put, that a superior machine intelligence could make decisions that result in harming or enslaving human beings.

This fear has been cultivated throughout history. One of our major world religions is based on the fear of forbidden knowledge in a garden. From Adam and Eve it's a short jump of several centuries to Mary Wollstonecraft Shelley, whose *Frankenstein* bears the alternate title of *The Modern Prometheus,* explicitly connecting the novel's artistic pedigree to the Greek myth of the figure who was punished for introducing humans to the knowledge of fire. In an interesting historical twist, one of Shelley's contemporaries was mathematician Ada Lovelace, often credited as the grandmother of computers. The dynamic between creative expressions of the fear of scientific progress and scientific progress itself was fortified in an actual human relationship between two Victorian women. And we still contend with the aftershocks of their revelations.

We're deeply ambivalent about quantum leaps in knowledge, and these fears and warnings make a certain amount of evolutionary sense.

Civilization has learned that scientific breakthroughs lead to unexpected consequences. Innovations threaten the status quo, and it can take a while for society to absorb new ideas. It's sane to be cautious.

We live in an era of accelerated change, which can lead to what my friend Steve Turnidge likes to call "the bends," the sense of disorientation and loss of control that can accompany rapid evolution. Our art gives expression to these anxieties. We can't get enough of dystopian scenarios of androids who turn the tables on their human creators and wastelands where punk rockers fight over oil. Our apocalyptic imaginations comfort us with tales in which a chosen few learn how to survive in a world gone mad.

The narrative that's coalescing around every new advance in artificial intelligence pits human vs. machine while neglecting to notice that we already live by the whims of a massive, decentralized super intelligence of our own creation. Our daily behavior and attitudes are governed by algorithms developed in Palo Alto and weaponized in Moscow. Trump's election and his subsequent dismantling of many of the institutions we take for granted as features of a civilized society (the EPA, the State Department, the NEA, etc. etc.) are a cognitive fissure between the twentieth and twenty-first centuries, a border separating old modes of outrage from the scandal-gorged, numb helplessness in which we're now miserably marinating. This new state could not exist outside the context of the artificial nervous system we use to exchange funny videos of goats dressed up as ducks.

When we express fear about artificial intelligence running amuck, we're projecting our intimacy with human evil onto our tools. We worry that machines that can improve their own intelligence will reach the point where they can be just the vile, greedy, heartless, and murderous bastards we've proven the human race can be. We're not afraid of machines becoming more human; we're afraid of machines retaining the violent, animal instincts that we humans have never been quite able to shake.

When sales of Orwell's *1984* spiked after Trump's election, we seemed to be reaching for an owner's manual of autocratic oppression. The book we should have been reaching for was Aldous Huxley's *Brave New World*, in which a society is kept blissfully distracted by sensation, drugs, and immersive feelies that uncannily resemble today's virtual worlds. My hunch is that if AI takes over, it won't resemble the future battlefields of the *Terminator* movies, in which human guerillas duke it out against Skynet's drones. The future that AI will deliver will be a future that we'll be manipulated into believing was our idea to begin with. We'll march willingly into luxury caves that have been designed for us as the robots assume dominion over the increasingly uninhabitable surface of the earth.

To what end? I have long suspected that the ultimate purpose of life on Earth is to seek and seed other life in the universe. Largely segregated from its own physical form, the human race will become the subconscious of a global intelligence so advanced as to be indistinguishable from God. The earth will extend itself ever outward as humans delve deeper inward, into the crust of the planet and the virtual habitats we'll plant down there. The AI will copy what we perceive in order to create backup copies of human beings that can simply be rebooted and run perpetually through endlessly branching life stories. Some even suggest that we exist in this state right now and we just don't realize it, chained to the wall in Plato's cave, longing to liberate ourselves into the blinding light.

## II. Will an AI Ever Learn to Love?

How does that question make you feel? Silly? Embarrassed? Stupid?

Love is the source of our greatest power but, paradoxically, is the thing that makes us feel most vulnerable, to the point that most of us avoid talking about it at all. We literally die without it when we're infants, and we organize our adulthoods around pursuing its cheap facsimiles. Its absence is at the center of our greatest mistakes and misfortunes. Finding other people to love is the primary project of most of our lives. Love is the cornerstone of every major religion and the subject of every brain-dead pop song. Love is simultaneously the most profound and frivolous element of human experience.

In the end, love is probably just synapses firing in a certain configuration in our brains, just like everything else that will be replicated by a quantum computer in the near future. And yet Cartesian explanations for love strike us as inadequate. Love feels like something bigger than what can be contained in the confines of a single human heart. Love feels ancient, a force that existed in the universe prior to the arrival of human beings.

When you start speculating about the nature of love, it's an easy hop from empiricism into the realm of Hallmark cards. Talking about love makes you look unserious, unless you're, say, the Beatles, who mused on the subject with unprecedented artistry. AIs are starting to write songs, too. Will they one day be able to sing about love in ways that move us to tears?

I wonder if the question of whether an AI can love is tied to more pragmatic questions about the very purpose of AI. Much of the cultural discussion surrounding AIs is about how many of our jobs they'll replace, or whether they'll destroy civilization through robotic insurrection. We look to a future of machines that can teach themselves how to learn and we feel a shiver of foreboding. I wonder if this foreboding is based less on fear of the power of computers than on the understanding that we'll have no choice but to confront our faults as a species. Our shortcomings will be exposed once and for all. How eerie are the echoes of the Singularity to the Christian concept of judgment day.

Richard Brautigan, a Bay Area poet most active in the seventies, once wrote a poem in which he referred to "machines of loving grace." Brautigan wasn't what you'd call a science fiction writer. He was more an absurdist who chased themes of belonging and romantic attachment like a lost dog pining for its owner. He excelled at jamming words or images together that you'd never expect to encounter side by side in a sentence or line of a poem. The line "machines of loving grace," written during the IBM era, is par for the course for Brautigan. Two ideas, technology and spiritual love, that surprise us by standing hand in hand.

Machines of loving grace are what we hope will evolve instead of the Skynet of the *Terminator* franchise. In our dystopian nightmares, we project onto our machines the most venal and destructive impulses of human nature. We assume that a machine as intelligent as we are will be as greedy as we are. We've plundered the resources of the planet to achieve a civilization capable of shattering subatomic particles and hacking DNA, and we assume that our all-consuming compulsion to propagate, expand, and conquer will be passed to our machine progeny. We hope that they'll have more mercy on us than we've shown to each other, to animals, and to the planet itself.

What's missing from this speculation and worry is a grand purpose for AI. The ultimate reason why billions of hairy bipeds evolved to create an entirely new kingdom of life. For years I've had a hunch about what that purpose is.

What if the purpose of technology is to spread life itself throughout the universe? Human beings exist within the context of nature and technology exists within the context of human invention, therefore technology is part of a natural process. But to what end? The clues are all around us. Planet earth wants to deliberately control its own physical, chemical, and biological processes to conquer the loneliness of fostering the only life that it knows. The unbearable loneliness of consciousness-infused matter requires said matter to organize itself in ever more ingenious ways to reach farther into the universe, to seed new living planets, to create more opportunities for consciousness to find a home in this particular universe. Perhaps the earth intends to propagate itself by sending spores beyond its boundaries.

The moment at which we can say that an AI has learned to love will be the moment we can no longer call it artificial.


III. I'm Pretty Sure I Could Teach a Machine to Write a Novel

There will come a day when we're enjoying novels and stories written by artificial intelligence. I've starting thinking about how this might actually happen. How you feel about this idea probably reveals your particular cultural orientation. If it horrifies you to imagine computers churning out works of fiction, then you're likely a citizen of the Gutenberg era, and you may be fuzzy on the growing capabilities of machine intelligence. If the notion of novels written by robots delights you, you might be up on the latest in AI but fuzzy on the messy, human process of creating art.

The first viable AI fiction will probably be the result of humans and machines working together. We won't wake up one morning to discover a computer has written a bestseller. We'll slowly merge machine intelligence and the human imagination to the point where these two things blur. In fact, we already are.

The trajectory of my writing life proceeds from pen and paper, to typewriter, to PC, to the auto fill selections in text messages and my gmail account. I've navigated this progression with varying degrees of discomfort and acceptance. I taught myself how to type and to this day just type with my thumbs and middle fingers. I adapted my writing process to machines, learned to depend on spell-check, have saved my stories on floppy discs and in the cloud.

In the mid-nineties, concerned about the incursion of technology into literature, I leaned heavily on the ideas articulated by Sven Birkerts in *The Gutenberg Elegies*, which argues that something vitally human is lost when we move from analog to digital modes of storytelling and communication. A big part of the reason I chose to enroll in Bennington College's MFA program was because Birkerts taught there. But Birkerts's passionate cries in the wilderness seem to have done little to slow the march of technology. I expressed my ambivalence on these issues in a story called "Written by Machines," which I wrote in 2001 as a wildly speculative scenario that strikes me today as too narrowly imagined. Nowadays, when I see the suggested responses that Google offers at the end of an email, a little part of me winces, especially when one of those responses is exactly what I would choose to say myself.

In coming decades, we will experience a wave of shocking abdications of activities we now assume can only be accomplished by flesh and blood humans. The world's chess masters and Alex Trebek long ago accepted that there are some intellectual tasks machines can perform faster and more accurately than humans. The history of technological development is a history of breaching lines we believe separate the human mind from the tools those minds have invented. The age of machine intelligence that is already well underway is about to shock us more than ever by colonizing the precincts of our imaginations.

Awhile back, my friend the novelist Christopher Robinson introduced me to the concept of the "centaur," a human-AI hybrid. The idea being that AIs won't completely replace the human creator, but that humans and AIs will work together. This makes sense to me as a novelist because that's already how novels get written, through the push-pull relationships with editors and, more broadly, a readership. Chris co-wrote his novels *War of the Encyclopaedists* and *Deliver Us* with his friend Gavin Kovite, so he's perhaps more open to the idea of two minds creating a work of art together than most. But every novel on your shelf, despite bearing the name of a single author, is the result of an editorial process that involves other people. The stories of Raymond Carver, it has been argued, owed as much to the scalpel of his editor, Gordon Lish, as to Carver's generous imagination.

Once you accept the notion of a story as collaborative exercise, maybe it's just a step to imagining a collaboration between a person and a machine. Isn't that what the author who pauses mid-draft to look something up on Wikipedia is already doing?

Recently, I had a conversation with Amber Osborne, an upstanding member of Seattle's VR/AR community who works as Chief Marketing Officer for Meshfire, a company that developed an artificial intelligence platform for social media. I wanted to get Amber's opinion on whether an AI could ever write a novel. She believes that this process is already underway, and mentioned a college student who developed an AI that writes his papers for him. Of course someone figured out how to do that.

But wait! My inner Sven Birkerts raises an alarm. Isn't the point of literature to express something irreducibly human? Go ahead and make that argument to all those kids who dig EDM and other genres of music generated as much by machines as by the humans who push the buttons. Human nature is under constant revision, and part of that revision is the way in which we internalize the nature of our tools.

I know a poet, a linguist, and a handful of experts on machine learning that I bet I could get into a room and hash out a process by which computers could start writing fiction. I'd be surprised if this wasn't already happening somewhere. What will truly shock us won't be when we're reading books written by artificial intelligence. It will be when AIs are reading them, too.

IV. The Highest Purpose of Artificial Intelligence is to Seek and
Propagate Life Throughout the Universe

Picture our moral imagination as a series of concentric circles. At
the center of this model is the self. As we grow, we generally move
outward from there. Our capacity for empathy guides us to invest
in the suffering of our immediate families, then our various tribes,
communities, and nations.

The moral innovators of our societies ask us to extend our empathy
to those whom we once believed didn't deserve it. Reverend Martin
Luther King Junior wasn't a great leader because he made conditions
better for African Americans. His greatness was the result of a moral
imagination that understood how civil rights improved the lives of
all Americans, regardless of race. The genius of Malala Yousafzai is
that her empathy extends to even the religious fundamentalists who
attempted to murder her. These thinkers challenge us to broaden the
categories of people we believe deserve our empathy.

Conversely, the world's villains are those whose moral imaginations
are selfishly restricted to themselves and a narrow band of people
they believe deserve empathy, tyrants who demonize "the other"' and
perpetuate the idea that "those people" don't deserve the empathy they
themselves demand. Gee, I just can't seem to think of any examples
of this off the top of my head.

Microsoft CEO Satya Nadella's book *Hit Refresh* recounts his
journey from India to the top leadership post of the Redmond technology giant, while promoting empathy as a guiding principle for the
development of artificial intelligence. Nadella writes:

> The challenge will be to define the grand, inspiring
> social purpose for which AI is destined... In 1961,
> when John F. Kennedy committed America to landing
> on the moon before the end of the decade, the goal
> was chosen in large part due to the immense techni-
> cal challenges it posed and the global collaboration it
> demanded. In similar fashion, we need to set a goal
> for AI that is sufficiently bold and ambitious, one that
> goes beyond anything that can be achieved through
> incremental improvements to current technology.

I humbly offer that this grand, inspiring goal could be to discover and propagate life throughout the universe.

Over the course of a typical human life, we find ourselves venturing further outside ourselves, empathizing with others who are not like ourselves. Our artists and moral innovators show us paths toward more expansive empathy for our fellow human beings. Usually we hit a limit.

When I saw the movie *Gandhi* with my father at age nine, one scene in particular rattled me and expanded my moral imagination. Mahatma Gandhi, played by Ben Kingsley, is comforting a Hindu man whose son has been murdered by Muslims. As an act of revenge, this Hindu man killed a Muslim child. The man is convinced he will go to Hell. Gandhi offers him a way out—find a child who has lost his parents, and raise him as his own. Most importantly, make sure this child is Muslim and raise him as a Muslim.

Here was an example of a moral genius challenging someone to broaden their empathy beyond the circle of their own faith. This seems to be the logical conclusion of humanity's great religious traditions. *Love thy enemy* is a radical challenge to elevate our morality, one to which I certainly can't seem to rise most days.

Nadella reveals the framework of his own moral imagination with the inclusion of a single word in the passage I quote above: "The challenge is to find the grand, inspiring social *purpose* for which AI is destined." (emphasis mine). What if we can find AI's purpose beyond the limiting sphere of human civilization? What if the only way forward for this profound technological leap is for it to be paired with a profound moral leap? What if the grand purpose of AI lies in the stewardship of life itself, in all its myriad forms?

Often, I find that people who make a living by developing technology fall into what I call the "smart fridge" trap. This is when the problems a technology fixes are disproportionately small to the potential of that technology. Many years ago, we started hearing about advances in tech that would lead us to the "smart home," in which everything in our domiciles was monitored for our convenience. I kept hearing variations on the following lame, utopian promise: just imagine, our refrigerators will automatically notify us when our milk goes sour!

Imagining what artificial intelligence will achieve only within the context of human society is just a larger-scale example of the smart fridge trap. To start grasping the grand purpose of AI, we must think way beyond the needs of our human species. If we expand our empathetic circle beyond Homo sapiens and embrace the responsibility of stewardship over all life—from microbes to fungi to earthworms to trees to gorillas—we can begin to see ourselves as a manifestation of earth's desire to spread life beyond our star.

Earth is a spore. Technology is a natural mechanism that the earth has developed, via a bunch of smelly primates, to propagate other planets. I explored this idea in a fictional framework in my novel *Blueprints of the Afterlife*, but I am increasingly convinced that this is ultimately the purpose for which we humans evolved. I've been obsessed with this idea in some form since I was a teenager, and as I've aged, the technology to actually act on this theory has increasingly come to pass.

On the wall of my living room is a chart titled "DEVELOPMENT OF LIFE" that illustrates the various geological ages of our planet. What's striking is how endlessly creative and resilient life has been on this far-flung rock, how agile life has been in recovering from numerous extinction events, culminating in a species that manipulates DNA, splits atoms, peers billions of light years beyond the sun, and laughs at its own farts. Even more striking is that in terms of geological time, humanity's rise has occurred in a crazy bang flash instant. My favorite explanation for how brief our tenure has been on earth is to imagine geological time to be as long as your arm. Then, imagine that a single swipe of a nail file across your middle finger's nail wipes out the entirety of human history.

The development of AI is another explosion within the explosion of humanity's rise to control and alter the very climate of the planet we occupy. Paradoxically, it might be the case that we had to imperil our planet to get to this point of technological development, and if this technology can reverse those changes, as Nadella briefly suggests, then maybe we'll get to stick around long enough to ask a bigger question: where do we go from here?

Perhaps we are heading toward an era in which humans will continue to exist, just not physically. We will exist as content within the AI as it traverses the universe, finding dormant planets to seed and cultivating fledgling life to grow to the point of sentience and macro-reproduction. Perhaps we already exist within such a realm and it's just beginning to dawn on us. Perhaps it will be artificial intelligence that charts the course as empathy takes us further from the nucleus of the self into the outer reaches of human society, from human life to all life, from life on earth to life beyond, and the beauty we'll find there.

# ACKNOWLEDGMENTS

The author wishes to thank Steve Turnidge, Paul Hubert, Elizabeth Scallon, Ron Jones, Christopher Robinson, Amanda Knox, Colin McArthur, Eric Reynolds, Dave Cornelius, Lisa Biagi, and his kids and parents for support and inspiration.

# ABOUT THE AUTHOR

Ryan Boudinot is the founder of Seattle City of Literature. He is the author of four books, including the novel *Blueprints of the Afterlife* and the story collection *The Octopus Rises*. He lives in Seattle.

CPSIA information can be obtained
at www.ICGtesting.com
Printed in the USA
FSHW04n0947180418
47138FS